Controlling Automated Manufacturing Systems

Controlling Automated Manufacturing Systems

PJ O'Grady

Kogan Page

First published 1986 by Kogan Page Ltd
120 Pentonville Road, London N1 9JN

British Library Cataloguing in Publication Data

O'Grady, P. J.
Controlling automated manufacturing systems.
—(New technology modular series)
1. Flexible manufacturing systems
I. Title II. Series
658.5'14 TS155.6

ISBN 1-85091-120-7

Printed and bound in Great Britain by
Biddles Ltd, Guildford and King's Lynn

Contents

Preface

This book is intended as an introduction to production planning and control of automated manufacturing systems. As such, it links together two diverse fields of interest: in the area of production planning and control there is a large body of work completed in analytical models, computer structures and overall systems; equally, for the hardware and detailed control aspects of the equipment used (for example, NC machines, robots, etc), comprehensive studies have also been completed. To cover each area fully would result in a work of several volumes. Instead, this book stresses the important elements of both areas that are vital to effective production planning and control of the whole automated manufacturing system.

Overall, the book presents a viable production planning and control structure for an automated manufacturing system. This structure has been designed to tie in, where possible, with existing more traditional production engineering and production management approaches, that may well already be firmly established within an organization. Detailed mathematical treatments have been avoided in favour of describing fundamental structures; but adequate references have been given for those eager to pursue more detailed aspects.

A strategy for the tremendously important area of production planning and control for automated manufacturing systems is provided. Feasible and effective approaches are described and their application and implementation is discussed.

Chapter 1 provides a brief introductory definition of an automated manufacturing system, and outlines a few of its most characteristic features and most significant areas of application.

Chapter 2 describes the particular requirements that automated manufacturing systems impose on the production planning and control system.

Chapter 3 explains the background against which a production planning and control system for automated manufacturing systems can be implemented. Included are brief overviews of master production scheduling, materials requirements planning, and job shop scheduling. Analytical and heuristic approaches that can be used to aid master production scheduling and job shop scheduling are reviewed.

In Chapter 4, the structure of the production planning and control system for an automated manufacturing system installation is given. In this book, the structure is divided into four hierarchical levels for production planning and control purposes. These four levels are the factory, shop, cell and equipment levels.

The following four chapters deal with each of these levels in turn, and Chapter 9 provides a conclusion to the book with a brief analysis of likely future trends in the development of automated manufacturing systems.

CHAPTER 1

Introduction

Rapid improvements in both computer hardware and software have made possible a dramatic shift towards automated manufacture. Complete mini factories can now operate with limited human involvement; such automated manufacturing systems rely on effective control procedures for their operation.

Conventionally, manufacturing can be divided into three major categories:

1. Flow or mass production
2. Batch manufacture
3. Jobbing manufacture

Flow or mass production is concerned with producing a limited range of products in high volume (for example, car assembly). Batch manufacture deals with a much larger product range than flow manufacture, but the products tend to have lower volumes and repeat orders are expected. Jobbing manufacture produces what may be termed 'one-offs', that is, there is no expectation that there will be repeat orders for the products. Jobbing manufacture is characterized by a high product-type range but a low volume.

In Western industrialized countries, the proportion of manufacturing output is greatest for batch manufacture; it is usually taken to be somewhere in the region of 70 per cent of total manufacturing output. This book therefore focuses most of its attention on batch manufacturing.

One frequently quoted aim of automated manufacturing systems is to raise the efficiency level of batch manufacture to the level of flow manufacture, and this can be greatly eased if an effective production planning and control system is used.

What is an automated manufacturing system?

A variety of terms have been used to describe highly automated manufacturing facilities, including:

Flexible manufacturing systems.
Computer integrated manufacturing systems.
Automated manufacturing systems.

Each of these terms, which tend to be used more or less interchangeably, describes a highly automated, integrated manufacturing facility. Purists may argue that the present generation of automated manufacturing systems are not particularly adaptable (see later) and should not therefore be labelled flexible manufacturing systems. These purists may also argue that a full computer integrated manufacturing system should include design, manufacturing, control and financial computer systems, and that automated manufacturing systems that do not contain all of these should not be labelled computer integrated manufacturing systems. Overall, therefore, it is perhaps safer to restrict the terminology to 'automated manufacturing systems' and this terminology is followed throughout the book.

Over the past few years a variety of attempts have been made to define an automated manufacturing system. A definition given by Draper Labs (1983) is perhaps a good starting point:

> 'a computer-controlled configuration of semi-independent work stations and a material handling system designed to efficiently manufacture more than one part number at low to medium volumes.'

The definition given by Groover (1980) is a more detailed one which gives some insight into the overall structure of automated manufacturing systems (although he does use the term FMS):

> 'An FMS consists of a group of processing stations (usually NC machines) connected together by an automated work part handling system. It operates as an integrated system under computer control. The FMS is capable of processing a variety of different part types simultaneously under NC program control at the various work stations . . . The work parts are loaded and unloaded at a central location in the FMS. Pallets are used to transfer work parts between machines. Once a part is loaded on to the handling system it is automatically routed to the particular work stations required in its processing. For each different work part type, the routeing may be different and the operations and the tooling required at each work

station will also differ. The co-ordination and control of the parts handling and processing activities is accomplished under command of the computer.'

This definition highlights the central role of the computer in co-ordinating and controlling the activities of the automated manufacturing system; this co-ordination and control function is fundamental to the overall automated manufacturing system efficiency.

Some other points arising from Groover's definition should be stressed. First, the simultaneous processing of a variety of part types is mentioned. This infers the careful co-ordination of different sections of the automated manufacturing system, so that part types can be passed from one section to another. Second, Groover indicates that each part type may have a different route, so planning and controlling the movement of a number of different part types through different routes may be a complex problem.

One frequently quoted aim of an automated manufacturing system for batch manufacture is to lower the cost of discrete part manufacture so that the cost more nearly resembles that of flow manufacture. This is achieved by several features of an automated manufacturing system:

1. Part programs can be downloaded to NC machines relatively easily.
2. Lead times can be reduced.
3. Levels of equipment usage can be raised.

The latter two features are, in particular, dependent on the provision of an adequate and effective production planning and control system.

WHAT IS FLEXIBILITY?

Automated manufacturing systems can perhaps achieve their greatest potential when they are designed to be flexible. This flexibility can take a number of forms, including:

(i) *Volume flexibility* – the ability to handle changes in the production volume of a part.
(ii) *Re-routeing flexibility* – the ability to have a number of routes through the system for each part in order to enable, for example, machine breakdowns to be dealt with.
(iii) *Part flexibility* – the ability to handle a wide variety of parts including the ability quickly to adapt the system to handle a new part.

Including this kind of flexibility into the design of an automated manufacturing system can greatly increase costs; it is perhaps not surprising that many modern automated manufacturing system installations are not particularly flexible. The Ingersoll Engineers' survey (1982) found that for the installations they investigated, compatible part numbers were, on average, restricted to eight and that the proportion of all components in the plant that passed through the automated manufacturing system was approximately four per cent. This fairly limited role for automated manufacturing systems is likely to change over the next few years as the second generation of system facilities come on stream. These second-generation automated manufacturing systems offer a much greater degree of flexibility in the production range.

Automated manufacturing systems have traditionally been associated with metal machining, thereby containing a number of direct numerical control (DNC) machine tools. Recently, much attention has been focused on other areas in which the concepts inherent in automated manufacturing systems can achieve major benefits. One such area is in electronic assembly, where a flexible assembly line can be built to handle a number of different assembly tasks. With good design of the flexible assembly line, new assembly tasks can be readily incorporated.

Why is production planning and control important?

Reduced lead times, low work-in-progress, low inventory levels and high facility usage are extremely important for *any* manufacturing system; consequently, production planning and control is important for all manufacturing systems. However, it becomes increasingly important in an automated manufacturing system for two major reasons. First, one of the significant advantages of automated manufacturing systems is that manufacturing lead times (ie the time taken to manufacture a part) can be shorter than in conventional manufacture. It is not unusual to find lead times for a part of six weeks in conventional manufacture being reduced to eight hours when that part is manufactured in an automated manufacturing system. These shortened lead times mean that the production planning and control function becomes much more important, since activities must be scheduled and controlled more closely to achieve these reduced lead times. Whereas in conventional manufacture, for example, waiting for specialized tooling can be tolerated, in an automated manufacturing system this delay is unacceptable; the specialized tooling has to be available as and when needed, otherwise lead times escalate.

The second reason for the importance of production planning and control in an automated manufacturing system is the high cost of most of these systems (several million dollars being typical), meaning that high system usage becomes an important factor. Most automated manufacturing systems probably aim at an average usage across the whole system of 85-95 per cent whereas 40-60 per cent is typical for conventional manufacture. The achievement of this high usage is, again, aided by an effective production planning and control system.

Effective production planning and control is tremendously difficult in both conventional manufacture and in automated manufacturing systems. This is because typical batch manufacture involves planning and controlling a large number of jobs through many machines/processes, posing an exceedingly difficult combinatorial problem.

The degree of difficulty in planning and controlling production in an automated manufacturing system is greater than that in conventional manufacture due to the two major factors indicated above. In particular, the requirements of short lead times (meaning that tooling, machines, transport and inspection must be available when needed), and the desire to achieve high usage of the automated manufacturing system, mean that there is a need for a much more sophisticated planning and control system.

CHAPTER 2

Automated Manufacturing Systems and Production Planning and Control

Introduction

The complexity of the problem of production planning and control for batch manufacture in both conventional and automated systems has been stressed: again, any automated manufacturing system will underachieve unless there is high quality production planning and control. The results of poor production planning and control can be severe, with high work-in-progress levels, high lead times and poor system usage. The latter aspect is important when one bears in mind the high cost of these automated manufacturing systems; the only way in which they can be justified is if they demonstrate the good rate of return associated with high usage.

This chapter gives some background to the special production planning and control characteristics that automated manufacturing systems require. The first aspect discussed is that data availability, quality and immediacy are fundamentally different from that found in conventional manufacturing systems; the greater accuracy in the data available to decision makers from automated manufacturing systems could therefore lead to better decisions.

There are, however, several other factors which can complicate the production planning and control system for automated manufacturing systems. These include: short manufacturing lead times; the consideration of engineering details; the greater emphasis placed on system usage; the need to integrate with existing software systems; and the need to generate detailed instructions. These factors are described in turn and their impact on the production planning and control function is discussed.

Factors affecting production planning and control

An automated manufacturing system differs considerably from its conventional counterpart. These differences lie not only in the manufacturing hardware, but also in the software and communication aspects, as well as in the provision of sensors and other monitoring devices. This means that automated manufacturing systems have the mechanisms to be closely monitored and controlled, so that production planning and control can be done with more certainty about the actual state of the manufacturing process than in conventional manufacturing systems.

Nearly all automated manufacturing systems contain some provision for communication within the system. In many cases this communication is achieved by direct linkage of elements of the system. In other cases a network will be used, so that the same transmission medium can be used for a variety of communications.

Overall, therefore, automated manufacturing systems are likely to produce a large amount of data that can be used for production planning and control. This data is likely to be produced automatically and is likely to have higher accuracy than that produced by conventional manufacturing systems. Furthermore, this data is likely to be up to date since there should be little elapsed time between the data being generated and it being received by the production planning and control systems.

Decisions made in such an environment can, therefore, be better for three reasons. First, a greater amount of data is available since it is generated automatically. Second, the data produced automatically is likely to be more accurate than that produced by human labour although, of course, a system should be protected against totally ridiculous data being produced by faulty equipment. Third, the data can reach the decision-making areas faster than in conventional systems.

Although decisions can be made under better conditions than in conventional systems, some features of automated manufacturing systems can result in more complex problems to be solved. These features are:

1. Manufacturing lead times are shorter.
2. Engineering details need to be considered.

3. Greater emphasis is placed on system usage.
4. The need to integrate with existing software systems.
5. Detailed instructions need to be generated.

SHORT MANUFACTURING LEAD TIMES

One of the major justifications for automating many conventional manufacturing systems is that manufacturing lead times will be considerably shortened. Achieving this has a number of consequences for production planning and control. First, there may well be a need for a second master production scheduling (MPS II) function (see Chapter 6). The master production scheduling I (MPS I) function in both conventional and automated manufacture is concerned with time-frames of (usually) a week, and it calculates the desired production rate for end-products for each week. This work load may then pass through a materials requirements planning (MRP) system to obtain a detailed work load, including component and raw material requirements. Since the automated manufacturing system has lead times which are desired to be low, this weekly work load may need to be filtered to extract a viable portion of work for perhaps only a few hours. This filtering can be done using an MPS II stage which is discussed in detail in Chapter 6.

Second, the speed at which jobs move from operation to operation means that a much more comprehensive on-line control facility is required. This facility must be capable of quickly and accurately determining the next moves to be made in the automated manufacturing system; there may therefore be a need for a faster response time from the production planning and control system than in the more conventional manufacturing systems. The problem of production planning and control in automated manufacturing systems can be broken down into a series of levels in a hierarchy of planning and control (see Chapter 4). In this hierarchy there are principally two modes of operation:

1. *Feedforward mode.* The commands flow down the hierarchy from the factory level, promoted by a change of time periods or a change in factory-level data. This mode is essentially a planning mode for the entire following time period and, as such, can be performed at a relatively leisurely pace. It is in this mode that perhaps the more analytical approaches may be useful.
2. *Feedback mode.* Some change in the plan is triggered by feedback

from lower levels, indicating that things are not going according to the schedule. The actual perturbations to the system can be evaluated and, if they are severe (for example, a machine breakdown), a new plan may have to be calculated in a very short timespan. Under these circumstances the more analytical approaches are likely to involve too much computation at present, and either a fixed heuristic or a look-up table may have to be used, although with future improvements in computer power then simple approaches to feedforward mode may well be possible.

Approaches that can be used for both modes are described in Chapters 5, 6, and 7.

ENGINEERING DETAILS

The low work-in-progress levels and low manufacturing lead times associated with automated manufacturing systems mean that much room for manoeuvre will be lost. With conventional manufacture, for example, considering the use of specialized tools when factory scheduling is not particularly crucial, since jobs can usually wait until such tooling is available. However, this delay is not desirable in an automated manufacturing system, and much thought on such aspects as tooling requirements and jig/fixture requirements may have to be included in the production planning and control procedure. Since the same tooling and/or jigs/fixtures may be required by more than one product type, the use of tooling and jigs/fixtures may have to be taken into account across the whole manufacturing system. In the hierarchy of such planning and control (presented in Chapter 4), the consideration of engineering details will have to be done at a high enough level to cover the whole manufacturing system. At this relatively high level, much attention will have to be given simultaneously to all the tooling, jig/fixture and product type combinations. These detailed engineering requirements can therefore considerably complicate the planning process. Methods for including these aspects are given in Chapters 5, 6, and 7.

SYSTEM USAGE

The aims of most production planning and control systems are to:

1. Achieve low throughput times.
2. Have low work-in-progress levels.
3. Achieve the job due dates.
4. Achieve high system usage.

In most circumstances these aims are conflicting: for example, achieving job due dates may mean that system usage, for parts of the system in any case, is low. The large capital cost of most automated manufacturing systems means that high system usage is usually thought to be of some priority when considering production runs. The desire to achieve high system usage but still to keep low throughput times, low work-in-progress levels and to meet job due dates is not an easy task, and can require a more sophisticated production planning and control system than that used in conventional manufacture.

THE NEED TO INTEGRATE

One major consideration in the design and development of a production planning and control system for an automated manufacturing system must be that it should integrate naturally with the existing software systems used by the organization. These existing software systems may form an integrated whole, particularly where they are from the same supplier. If this is the case, there are very probably readily useable interfaces between the software systems. On the other hand, the existing software systems may form a rather diverse group, especially where they have been bought from different suppliers or written in-house. If this is so, then suitable interfacing may be more difficult. Whatever the case, whether the systems form an integrated whole or whether they are more diverse, it is unlikely that an industrial concern will be willing to dispose of all the existing software packages. The automated manufacturing system's production planning and control software should therefore link in readily with the organization's existing software packages (such as materials requirements planning, computer aided design, and process planning). This is an important factor when designing the planning and control structures (see Chapter 6).

THE NEED FOR DETAILED INSTRUCTIONS

Human workers, particularly the more skilled ones, do not need to be given detailed instructions. Often, a broad outline 'goal' is all that needs to be given. For example, a skilled lathe turner will frequently need only to be given the finished dimensions of a part. Often, he will then select the base bar stock to be used, the tooling necessary and the speeds/feeds to use when machining. Furthermore, he will also sequence operations at the lathe to finish the part. For automated manufacturing systems, however, all the detailed instructions have to be generated within the production planning and control/process planning system. Furthermore, the activities have to be co-ordinated and scheduled in great detail. The provision of detailed instructions occurs mainly at the lower levels of the hierarchy presented in Chapter 4.

Conclusion

This chapter has described some of the features that make production planning and control of automated manufacturing systems different from that of conventional manufacturing systems. The first feature mentioned was data. The data present in automated manufacturing systems is likely to differ from that of conventional manufacturing systems in three respects:

1. *Quantity*. A greater amount will be available since it is generated automatically.
2. *Quality*. The data is likely to be more accurate, although of course a faulty sensor, etc could lead to dramatic errors.
3. *Immediacy*. There will be minimal delays between the data being generated and it being made available to the decision-making areas. In comparison, considerable delays can accrue between the data being generated and it being made available to decision makers with a conventional manufacturing system.

These three differences mean that decision making can probably be done with more certainty in an automated manufacturing system than in a conventional manufacturing system. However, this chapter has also pointed to some factors which will tend to increase the complexity of production planning and control for automated manufacturing systems. These factors include:

1. Short manufacturing lead times, leading to a requirement for faster decision making.
2. Engineering details need to be considered, thereby considerably complicating the procedure.
3. Greater emphasis is placed on system usage; the simultaneous achievement of this and other conflicting requirements, could prove difficult.
4. The need to integrate the production planning and control system with existing software systems.

5. Detailed instructions need to be generated for automated equipment and some provision for this has to be made.

Overall, therefore, although the problem of production planning and control has been aided by the quantity, quality and shorter elapsed time of the available data, it has also been considerably complicated by the above factors. The structure and approaches described in this book have been designed to operate within the constraints outlined in this chapter.

CHAPTER 3

Traditional Production Planning and Control

Introduction

A conventional batch manufacturing facility consists of a number of manually operated machines; jobs flow through the facility, following a number of different routes, with each job being processed on one or more machines. For most batch manufacturing concerns, perhaps the major characteristic that affects the production planning and control function is the complexity of the manufacturing system. A typical batch manufacturer may have thousands of different batches each month passing through a hundred machines. The interrelationships between jobs flowing through the manufacturing systems leads to a tremendously intricate combinatorial problem. The traditional approach that has met with some success has been to break down the problem into a hierarchy of levels.

This chapter gives an overview of this traditional approach to production planning and control in conventional manufacturing systems. First, the different levels of the hierarchy are reviewed. Traditionally, three levels are included, which coincide with a shortening of the planning horizon as the hierarchy is descended. These levels are long term, medium term, and short term. The long term level considers the overall manufacturing concerns with a planning horizon measured in years, while the short term level deals with units of time measured in minutes.

The medium term and short term levels are of most interest when considering production planning and control, as they deal with a planning horizon usually lying within a few months. The medium term planning stage is then discussed and some analytical methods of determining a medium term plan are given.

Assembly-type industries require the breakdown of the end-product medium term plan (often called the master production schedule (MPS)) into requirements for components, parts, materials and sub-assemblies; this can be done using a materials requirements planning (MRP) system. However, many MRPs fail to perform well in practice, and some reasons for this are given.

The last section of this chapter deals with the short term planning stage (often called job shop scheduling), where the jobs are sequenced at each machine. The most straightforward approach used is that of fixed heuristics, operating on a queue of jobs at a machine. The operation of these fixed heuristics is discussed.

Planning hierarchy

Batch manufacturing is usually concerned with manufacturing a wide range of product types and can involve considerable complexity in production planning and control. The approach that has been used with some success, has been to break down the problem into a series of levels in a hierarchy of planning and control. The most common arrangement is for these to be three levels, as follows:

1. *Long term plan* (sometimes called the corporate game plan: typically 0-5 years). This plan is a strategic examination of the position that the manufacturing concern occupies in the market. It considers such major aspects as the development of new products, the procurement of new plant, and changes in the workforce level in order to arrive at an acceptable scenario for the concern over the next few years. Typically, this plan is updated at quarterly intervals and involves participants from senior levels with the organization.
2. *Medium-term plan* (sometimes called the master production schedule (MPS): typically 0-1 years). The medium term plan or MPS considers the output from the long term plan as well as the expected demand, inventory levels and capacity levels, to produce a viable work load in terms of end-product production. This is often in weekly time periods over (typically) the next year. The output of the MPS is the volume of each product to be produced in each week. This MPS stage is designated MPS I in later chapters.
3. *Short term plan* (sometimes called detailed or job shop scheduling: typically 0-1 week). The output from the MPS stage consists of a relatively feasible weekly work load. For assembly-type industries, this work load can then be broken down into requirements for parts, sub-assemblies and raw materials, using an MRP system. The output from the MRP system (or the output from the MPS for non-assembly type industries) can then be fed into a short term planning or job shop scheduling stage. The function of this stage

includes capacity loading and sequencing each job on each machine or process.

Again, production planning and control is mainly concerned with a planning horizon of less than a year; consequently the two levels that are of most concern to production planning and control are the medium term (or MPS) level and the short term (or job shop scheduling) level. These two levels are now discussed in more depth.

Master production scheduling (MPS): medium term

The MPS stage considers a number of overall factors in determining a viable and effective MPS. These factors include:

1. Stock levels.
2. Forecast of demand (for make-to-stock operations).
3. Orders (for make-to-order operations).
4. Capacity levels.

The process of determining the MPS involves calculating the overall requirement for end-product production, using the forecast of demand and/or the orders received and subtracting finished product stock levels. This overall requirement is then adjusted in the light of the capacity levels available and other restraints, to result in a feasible amount of work to be done over the (usually) week considered.

The determination of the MPS involves participation from a number of different areas within the organization, including production, finance, sales, and engineering.

The view taken of a particular MPS is likely to vary from area to area within the manufacturing concern:

Production areas would like the MPS to give very *smooth work loads;*
Finance areas would like the MPS to give *low stock levels* so as to reduce the cost of supporting the capital tied up (make-to-order operations);
Sales areas would like the MPS to give *high stock levels* in order to increase the likelihood that an unexpected customer order can be met (make-to-forecast operations);
Engineering areas would like to have very *long lead times* made available to them for the design of customer specific products. This is of major importance in the make-to-order operations.

As a result of these sometimes conflicting requirements for the MPS it is not surprising that most successful MPS operations usually involve input from high level representatives of several functions within the manufacturing concern; especially from the marketing, finance and production areas. With these representatives there is then a better chance of the MPS being able to meet a more balanced mixture of requirements from different areas.

For the MPS function to operate effectively there are a number of other aspects to be considered:

1. The use of capacity data. Theoretical capacity figures obtained from, perhaps, work study sources are not usually a good indication of the likely future performance of a manufacturing system. More likely to give such an indication is the use of historical output figures. Such figures need some care in interpretation, making sure to allow for new factors such as product mix changes and/or new plant procurements.
2. The process of determining an effective MPS relies on senior level representatives from a variety of areas within the organization being able rapidly to appreciate the effects of a particular MPS. The inclusion of large amounts of data can make this process difficult; hence it is usual to reduce the number of different products considered, especially where the organization manufactures a large number of products. This can be done most effectively by grouping products together into product families with generically similar production process requirements. The number of product families that should be included in an MPS in order still to retain ease of comprehension, should probably be as low as possible although high enough to indicate the full effects of the particular MPS.
3. The units of measurement that are used in compiling the MPS should be kept uniform across the product range. Moreover the units should be capable of being easily interpreted in the light of the requirements of the manufacturing concern. Consequently, it is usual, for example, for capacity requirements for each product to be measured in hours with the total capacity requirement being the addition of all the sub-requirements. The stock levels are usually measured in units of the home currency in order to aid financial considerations.

The MPS function is a very important stage in production planning and control. What is also perhaps of major significance in the manufacturing organization is that it forces areas that might otherwise operate almost entirely separately (for example, production, sales and engineering), to communicate and discuss their real requirements on a regular basis.

ANALYTICAL APPROACHES TO MASTER PRODUCTION SCHEDULING

The process of determining a master production schedule (MPS) can be aided by the use of a suitable analytical approach. A wide variety of such analytical approaches are possible including:

1. Linear programming and integer programming.
2. Queueing theory models.
3. Linear decision rules.
4. Switching heuristics.

Of the above approaches, linear programming and integer programming (LP and IP) have probably gained the widest attention from researchers, although the number of actual applications has been relatively low. LP approaches rely on the expression of the constraints on the system as a series of linear equations. For example, if the total production of product X and product Y in period i must be less than (or equal to) 5200 units then this can be written as:

$$x_i + y_i \leqslant 5200$$

where x_i is the production of X in period i and y_i is the production of Y in period i.

The performance of the solution is evaluated as a linear equation and computer software packages can be used to optimize the solution in order to maximize (or minimize) the performance equation. IP concepts can be included when it is necessary to specify that certain variables can only take integer values, although this can increase computation time considerably.

Advantages of the LP and IP approaches are that the models are relatively simple to understand and that constraints can be readily incorporated. The main disadvantages are the necessary approximation to a linear function and the assumption that the whole of the operation is deterministic. Gunther's (1981) results suggest that LP models perform worse than some other models under stochastic conditions. In traditional batch manufacturing concerns, one other major problem in using LP and IP formulation is the size and complexity of the model produced, with hundreds of constraints being the norm.

The modelling of manufacturing systems as a queueing network using queueing theory to obtain solutions has been considered by several researchers including Solberg (1976), Buzacott and Shanthikumar (1980), and Yao (1983). The advantage of using queueing theory is the relative ease in obtaining solutions and it may also be useful for rough-and-ready calculations, although there are some rather severe assumptions used in queueing theory derivations.

The use of the HMMS linear decision rule (LDR) (Holt *et al.*, 1960) can also aid MPS. The original HMMS model relies on the expression of the product associated costs in a fixed quadratic format:

$$\sum_{t=1}^{T} [(C_1 - C_6) W_t + C_2 (W_t - W_{t-1} - C_{11})^2 + C_3 (P_t - C_4 W_t)^2 + C_5 P_t + C_{12} P_t W_t + C_7 (I_t - C_8 - C_9 S_t)^2 + C_{13}]$$

where $C_1, \ldots\ldots. C_{13}$ are constants for the particular system; T is the number of time periods considered; W_t is workforce level for period t; P_t is production rate for period t; S_t is sales in period t; I_t is inventory in period t.

The HMMS model is subject to the usual inventory constraint:

$$I_t = I_{t-1} + P_t - S_t$$

When the costs are minimized over a long planning horizon then the result is an LDR when the optimum values of the production rate P_1^* and of workforce level W_1^* are given by:

$$P_1^* = k_1 + k_2 I_0 + k_3 W_0 + \sum_{i=1}^{12} k_{i+3} S_i$$

$$W_1^* = k_{16} + k_{17} W_0 + k_{18} W_0 + \sum_{i=1}^{12} k_{i+18} S_i$$

where $k_1, \ldots, k_{30}$ are coefficients chosen in order to minimize the quadratic cost function above.

As can be seen, both the optimum production rate P_1^* and the optimum workforce level W_1^* are linear combinations of the inventory and workforce levels in the previous time period (period 0), and of the expected sales in periods

1 to 12. The forecast of future sales therefore falls naturally into the solution. Using the HMMS LDR, the optimum values of $k_1, \ldots k_{30}$ are calculated every time there is a substantial change in the cost function and then these values of k_i can be used, on a simple calculator on a regular basis, until there is a change in the cost function.

The original HMMS model is of a simple single product system with no manufacturing delay. Advantages of the approach are that the solution is relatively simple, the forecast of demand is included and it is capable of handling stochastic elements. However, critics of the approach do stress the relatively inflexible cost structure and the very simple model used (extensions partially to overcome these drawbacks have been given by Bergstrom and Smith (1970), Chang and Jones (1970), Welam (1975), and O'Grady (1981)).

Switching algorithm approaches (Orr, 1962; Elmaleh and Eilon, 1974) assume that production can only be carried out at discrete levels. When the inventory level is low, production is started and when the inventory level is high, production is stopped. Variants are possible whereby intermediate inventory levels trigger a change in the production volume. The simplicity of the approach is very attractive to practitioners but there can be problems in operation in a multi-product environment since capacity usage is not implicit in the approach.

O'Grady and Byrne (1985) have produced a variation on these switching algorithms by calculating what is termed the net excess stock (NES), ie the expected stock remaining at the end of the manufacturing lead time, taking into account the forecast of demand. A priority list is then produced by placing the products in increasing order of NES, on the basis that the items with the lowest values of NES are the products which are in most danger of having stockouts. Production is then scheduled by going through the priority list sequentially until the capacity levels are reached. The value of net excess stock Si(N), in period N for product i, is calculated as follows:

$$S_i(N) = C_i \sum_{J=1}^{L_i} U_i(N-J) - \sum_{J=0}^{L_i - 1} F_i(N+J) + I_i(N-1) - G_i$$

where for product i: $S_i(N)$ is 'net excess stock' at period N; L_i is expected manufacturing lead time; $U_i(N)$ is the quantity scheduled on to the manufacturing systems for period N; $I_i(N)$ is finished stock level at end of period i; C_i is expected pass rate of the manufacturing process (where pass rate = 1 — scrap rate); $F_i(N)$ is forecast of demand for period N; G_i is safety stock.

Variations in the approach are possible to take into account, for example, sequence dependent set-up times. Industrial studies undertaken on conventional manufacturing systems indicate that the approach works well.

The above are therefore some approaches which can be used as aids in selecting a suitable MPS. As indicated, the approaches each have their particular advantages and disadvantages. What is likely to happen in practice is that the relatively simple approaches can make significant improvements in the MPS, but that the law of diminishing returns applies to the more sophisticated approaches, in that they may only lead to relatively minor further improvement.

Materials requirements planning (MRP)

For assembly-type industries, the output from the MPS function is usually a viable work load of end-product production which can then be broken down into requirements for parts, sub-assemblies and raw materials using an MRP system (see Orlicky (1975) for a description). The MRP system is computer based and has the product structure in a computer file called the bill of materials (BOM). Lead times are also kept in a file. MRP uses the end-product demand, the BOM and lead time data to give the gross requirements for the components, together with the timing data necessary for correct planning. The net requirements are obtained by subtracting the components on order, in stock or in process.

Although MRP systems involve relatively simple calculations, in practice many MRP implementations fail to live up to expectations. There are a number of reasons for this:

1. *Poor MPS.* The MPS acts as a driver into an MRP system. Any errors in the MPS result in inaccurate output from the MRP system.
2. *Poor stock level recording.* The stock levels of particular components

are used to calculate the net requirements. Errors in stock level recording result in errors in net requirements.
3. *Inaccurate lead times.* The lead times are used as a fixed entity to give the time-phasing inherent to MRP operation. In practice, lead times are likely to vary and the use of a fixed lead time may result in difficulties, although Orlicky (1975) stresses the use of variable priority codes to ensure that lead times are less variable.
4. *Inaccurate bill of materials.* The enormous task of inputting the BOM file for most batch manufacturers usually means that some errors in the file are likely to occur. Although these start-up errors are likely to be rectified over a period of time, the necessity of frequent changes to the BOM in a fast-changing technological environment often means that errors occur on a continual basis.
5. *Out-of-date information.* In many MRP implementations data is recorded on paper and then entered manually via a keyboard into the computer files. The information that is on the computer may therefore be somewhat out of date. This problem can be reduced by the use of an effective shop floor data recording system (SFDRS). Usually an SFDRS relies on the use of a wand to read either bar codes or magnetic stripes and thereby to enter data directly into the computer from the shop floor.

Overall, therefore, although the MRP software itself may be working well, there may be problems with other aspects including data accuracy, that will cause difficulties in MRP implementations.

Job shop scheduling: short term

The detailed weekly work load from the MPS and also, in assembly-type industries, from the MRP system, can be scheduled on to each machine or process using a procedure called job shop scheduling. However, to obtain a good sequence is again a complex combinatorial problem. In relatively few circumstances, a good sequence can be obtained by evaluating all possible sequences, although the computation required rises dramatically with the size of the problem. For example, to evaluate all the sequences that are possible when 60 jobs are waiting to be processed at one machine would take 2×10^{68} years, even assuming that a million sequences could be evaluated each second! Therefore, although evaluation is possible for extremely simple problems, the computation inherent in larger problems renders the approach ineffective.

The problem of sequencing jobs has received a great deal of attention from researchers and practitioners. The approach that has perhaps been the most successful in practice is that of using heuristics operating on the jobs queueing at each machine or process. The way in which these operate is to rank the jobs in each queue on the basis of some simple measure and, when the machine or process becomes available, to choose the job at the top of the ranking. One simple measure is the ranking of jobs in order of the next operation time, with the job with the shortest operation time having the highest rank. This is called the shortest processing time (SPT) rule and has proved in practice to be extremely effective in reducing average throughput time. However, two major disadvantages of the SPT rule are that:

1. No account is taken of the required due date of the particular job. It is of little benefit to rush a job with low operation time through the manufacturing system if its due date is some time away and other more urgent jobs with slightly longer operation times are kept waiting.
2. Jobs with relatively long operation times rank low in priority whereas these jobs, with high added value, may well be particularly profitable for the organization. The shortest operation time jobs, which rank high using the SPT rule, may well be of low profitability.

Many other sequencing heuristics have been proposed including:

1. *Slack sequencing*. Due date slack is the difference between the due date and the total operation time for a job. The greater the slack, the greater the free time available for the job. A job with negative due date slack is in grave danger of overrunning the due date. A simple heuristic is to sequence jobs on the basis that the job with the smallest slack is sequenced first.
2. *Due date sequencing*. Jobs with the nearest due date are sequenced first. Note that using this heuristic, no allowance is made for the expected operation time.

The heuristics outlined above are generally fixed and do not alter or adapt to a particular manufacturing system. Therefore, as such they give good performance only in a narrow environment. As will be discussed later, the requirements for automated manufacturing systems are for much more adaptable heuristic approaches, and details of these will be given.

Conclusion

This chapter has discussed traditional production planning and control in batch manufacturing industries. A typical batch manufacturing system may well involve thousands of jobs passing through a shop floor containing a hundred machines, and the problem of adequately planning and controlling the progress of the jobs is tremendously complex. The approach that has met with some success has been to break the problem into a number of levels in a hierarchy of planning and control. The most common number of levels is three: long term, medium term (or MPS) and short term (or job shop scheduling).

The medium term (or MPS) level has been described and the necessary input from a wide selection of areas within the manufacturing concern has been stressed. Analytical methods for determining a viable and effective MPS have been described, as well as the advantages and disadvantages of each.

For assembly-type industries, it is often desirable that the output from the MPS is placed into an MRP system to obtain requirements for parts and sub-assemblies. However, in practice many MRP systems fail to live up to expectations and some reasons for this have been given.

The short term level (or job shop scheduling) involves the detailed sequencing of jobs on a machine and this can be a prohibitively difficult combinatorial to resolve. One approach that works well is to give a priority to each job in a queue on the basis of some fixed heuristic, the most common of which is the shortest processing time. Some other fixed heuristics rules have been described.

CHAPTER 4

Production Planning and Control Structure for Automated Manufacturing Systems

Introduction

The previous chapters have indicated some important characteristics of production planning and control systems. The first is that the problem of production planning and control for batch manufacturing is tremendously complex; a wide variety of interrelationships exists within a typical manufacturing system. This problem can be approached by using a hierarchy of control where major aggregate decisions are made at the highest levels and these decisions are gradually broken down into more detail as they pass through to the lower levels. The top levels of the hierarchy are therefore concerned with broad aspects of decision making, ensuring that the manufacturing system is globally achieving the desired objectives. Little detail is used at these higher levels but instead, such entities as aggregate production rates and capacities are used. The time periods considered tend to be longer at these higher levels, decreasing as the hierarchy is descended; where time periods measured in years may be used at the highest levels, the time periods may only be minutes at the lower levels. Overall, therefore, the characteristic complexity of the production planning and control task for manufacturing systems can be approached via hierarchical planning and control. As we descend the hierarchy the level of detail increases, whereas the time period considered decreases.

The second characteristic is that production planning and control is substantially different in automated manufacturing systems than in conventional manufacturing systems. This difference arises from three fundamental factors:

1. The capital investment in an automated manufacturing system is very high.
2. The control of an automated manufacturing system is usually computer oriented.
3. Manufacturing lead times are low.

These three factors have a number of repercussions. First, because the capital investment in expensive equipment is high, a greater emphasis has to be placed on ensuring that there is high usage of the system. Second, since control is computer oriented, there is more of an impetus to integrate the software with costly existing software systems within the organization, in order to achieve efficient flow through design, process planning and manufacture. Third, the reduced manufacturing lead times for automated manufacturing systems mean that decision cycle times will tend to be shorter. This may well lead to a heavier reliance on computer based decision support systems to produce effective decisions within the short time-scales permitted. Another factor associated with short lead times lies in the provision of manufacturing resources: whereas, in conventional manufacturing systems, waiting for tooling, etc can be accommodated within the relatively long lead times, the short lead times of automated manufacturing systems mean that engineering details (such as tooling and other manufacturing resources) have to be considered to ensure that these resources are available as and when necessary.

The fact that production planning and control is substantially different in automated and conventional manufacturing systems leads to, therefore, a number of repurcussions; these include a greater emphasis on system usage, the need to integrate with costly existing software systems, the provision of computer-based decision support systems, and the consideration of engineering details such as tooling and other resources.

This chapter describes a hierarchical approach to control of automated manufacturing systems, first giving a broad overview of two generic hierarchical approaches to this control: these being the advanced factory management system of Computer Aided Manufacturing Inc, and the advanced manufacturing research facility of the National Bureau of Standards. The second part of the chapter is concerned with extracting the essential elements from these two approaches to give a broad-based hierarchical control for automated manufacturing systems.

Advanced factory management system

Computer Aided Manufacturing International Inc (CAM-I) is an organization dealing with the design and implementation of computer technology in manufacturing. It operates on a consortium basis; individual companies join and then pay a further subscription to join one or more of a number of programs which carry out activities in a particular area. One particular program is the factory management program which is concerned with the design and implementation of a factory management system to manage production efficiently.

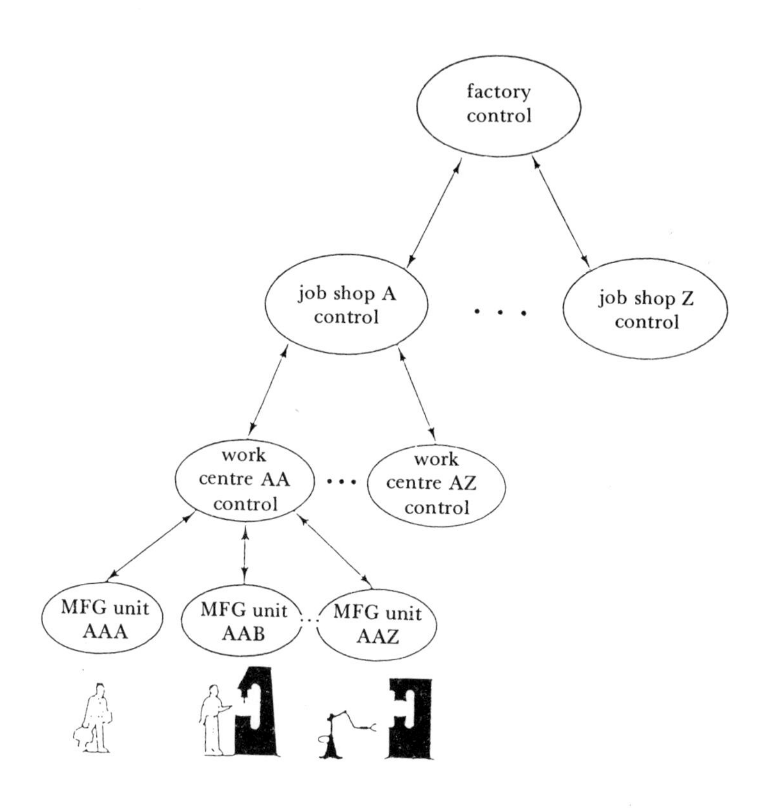

Figure 4.1 *Advanced factory management system hierarchy*

The factory management program has designed a computer hierarchy of control (see Figure 4.1) to break down the complex problem of planning and controlling shop floor activities into a series of smaller modules. The hierarchy consists of four levels:

1. *Factory control system.* This is the top level and concerns top level factory management. It considers such aspects as determining end-item requirements, product structure definitions (process planning), and individual shop capacities and capabilities.
2. *Job shop level.* This level is directly below the factory level. It takes commands from the factory level in order to determine commands for the work centre levels. Included in this would be the taking of end-item production and exploding this into processing operations. Having done this, the shop order events are scheduled.
3. *Work centre level.* This level takes commands from the job shop level and generates detailed task requirements. The task events are then scheduled and commands for these tasks are passed to the next level – unit/resource level.
4. *Unit/resource level.* The tasks from the work centre level are broken into subtasks and these subtasks are carried out.

Each level also has an associated feedback mechanism, whereby the occurrence of events is fed back to the level directly above. This procedure leads to a relatively decentralized structure, with decision making made at the lowest possible level commensurate with overall efficiency. Control of each level resides in the next highest level and this level issues commands to the level below. This lower level gives feedback on its current status to the level above, in order to facilitate decision making at that higher level. As we progress down the hierarchy the planning horizon shortens. At the top factory level, planning horizons may be of perhaps months, whereas at the unit/resource level the planning horizon may only be measured in seconds or minutes.

The events occurring at each level can be generalized. They are:

1. Model resources.
2. Maintain planning information.
3. Generate requirements.
4. Determine timing.
5. Plan resources.
6. Initiate events.
7. Monitor status.
8. Predict events.
9. Evaluate performance.

The particular events occurring at each level are shown in Figure 4.2.

The present developmental status of the CAM-I AFMS is that a comprehensive data flow model has been completed, as has a data dictionary. No implementation of these models has yet been carried out and this stage awaits further developments.

Automated manufacturing research facility

The National Bureau of Standards (NBS) are implementing a research and development facility for automated manufacturing. This facility, called the automated manufacturing research facility (AMRF), was originally envisioned as a testbed for evaluating automated metrology but has grown to include the development and testing of interface standards for future factories.

The first stage of hardware implementation consisted of two machining workstations (each with storage, a robot and a numerically controlled machine) together with a material transport system (with an automated guided vehicle). A major emphasis in the development of the control software has been to integrate the workstations in a manner which allows flexibility in the system configuration.

As with the CAM-I AFMS, the complex planning and control problems inherent in the AMRF have been broken down into a series of levels in a planning and control hierarchy, as shown in Figure 4.3. (see Jones and McLean, 1984; Furlani *et al.*, 1983).

At the top is the *facility* level, which includes process planning, production management (including long term schedules), and information management (including links to financial and other administrative functions).

Below this is the *shop* level, which manages the co-ordination of resources and jobs on the shop floor. The processes involved at this level include the grouping of jobs into part batches using a group technology (GT) classification scheme. The concept of a virtual manufacturing cell is introduced at this stage. These virtual manufacturing cells comprise machines which are grouped together in a dynamic fashion, ie the configuration and number of virtual manufacturing cells varies with time. This virtual manufacturing cell concept is further discussed later. Besides job

	MODEL RESOURCES	*MAINTAIN PLANNING INFORMATION*	*GENERATE REQUIREMENTS*	*DETERMINE TIMING*
FACTORY LEVEL	determine shop capacity and capability	update product structure definition	establish end item production	schedule end item production events
JOB SHOP LEVEL	determine work centre capacity and capability	update process routeing information	explode item requirements into processing operations (create shop orders)	schedule shop order events
WORK CENTRE LEVEL	determine resource capacity and capability	update process description	explode operations into detail tasks	schedule task events
UNIT/ RESOURCE LEVEL	verify capacity to perform	update control information	translate tasks into subtasks	schedule subtask events

Figure 4.2 *Automated factory management system control structure summary*

PLAN RESOURCES	*INITIATE EVENTS*	*MONITOR STATUS*	*PREDICT EVENTS*	*EVALUATE PERFORMANCES*
configure shops and plan production for shops/ suppliers	set end item production requirements and requests	monitor actual and predicted completion of items	predict completion of products	report product manufacturing performance
configure work centres and plan operations for work centres suppliers	release shop orders (order processing)	monitor actual and predicted completion of operations	predict completion of parts	report manufacturing performance
configure units and plan tasks for resources	despatch work assignments to resources	monitor actual and predicted completion of task	predict completion of operations	report process operations performance
allocate resources to subtasks	perform process (subtasks)	monitor actual and predicted completion of subtasks	predict completion of tasks	report task performance

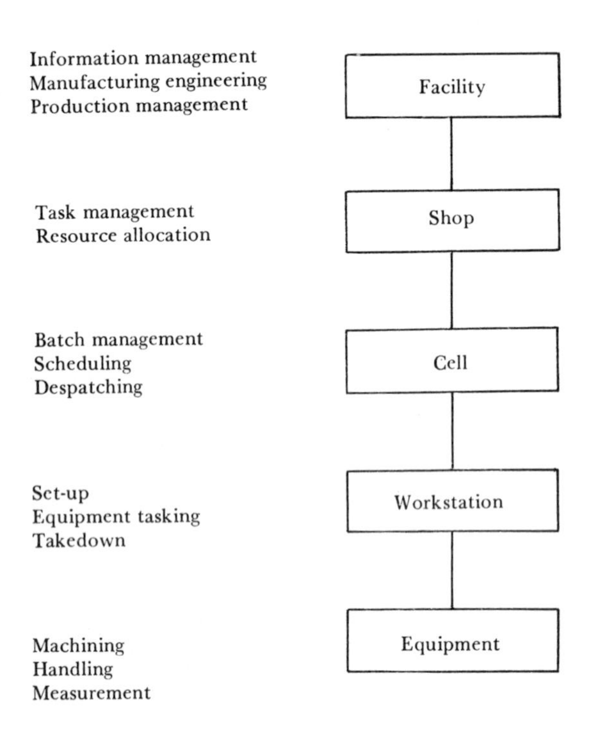

Figure 4.3 *Automated factory management system control hierarchy*

groups and virtual manufacturing cell configuration, the tasks at this shop level include allocating tooling, jigs/fixtures and materials to specific workstation/job combinations. These activities at the shop level are re-evaluated on the basis of feedback from the cell level and on changes in requirements, etc from the facility level.

Below the shop level is the *cell* level, where the cell controls system schedules the jobs. These jobs have already been divided into groups, with the jobs allocated to each cell being somewhat similar. Also involved in scheduling and controlling the jobs is the scheduling of material handling and tooling within the cell.

The level below is the *workstation* level, which consists of co-ordinating the activities of the AMRF workstation which is taken typically to consist of a robot, a machine tool, a material storage buffer and a control computer. The workstation controller then arranges the sequencing of operations in order to complete the jobs allocated to the cell control system.

The lowest level of the planning and control hierarchy is the *equipment* level, which consists of the controller for individual resources such as machine tools, robots or material handlers.

VIRTUAL MANUFACTURING CELLS

The concept of a virtual manufacturing cell has been proposed by NBS. This concept is somewhat similar to the traditional group technology approach, in that jobs are grouped into families whereby all the jobs within a family have similar manufacturing requirements. The major differences between virtual manufacturing cells and group technology is in the *dynamic* nature of the virtual manufacturing cell: whereas the physical location and identity of the traditional group technology cell is fixed, the virtual manufacturing cell is not fixed and will vary with requirements.

The impetus to develop the virtual manufacturing cell concept was based on the belief that the needs of the factory in the future could be met by a hierarchical structure that could dynamically alter its subsystem allocations as requirements dictated. This 'virtual' cell, therefore, is not associated with fixed groups of machines but individual workstations are allocated to it full-time or on a time-sharing basis with other virtual manufacturing cells. When requirements alter, the allocation of individual workstations will change, so that the virtual manufacturing cell is not identifiable with a particular set of workstations but is now a dynamically changing set of workstations.

However, to accomplish the implementation of the virtual manufacturing cell concept, two major developments have to be made. First, an increase in cell intelligence is required to handle the dynamic allocation procedure, including the possible time-sharing of workstations with other cell controllers. The second major development is that of ensuring that command and control communications, protocols, and handshaking are sufficient to meet the requirements. The scenario for a virtual manufacturing cell implementation includes the dynamic reassignment of workstations to cell controllers. Such a reassignment would mean that handshaking and reallocation procedures would have to be well developed.

Other problems may occur in the time-sharing aspect of virtual manufacturing cells, where workstations are allocated to cells on a time-slice basis. It is envisaged (McLean *et al.*, 1982) that there would be an interrupt point at which a part would be allowed to be removed from a machine tool prior to its completion on that machine tool, so that the workstation could be reallocated to another cell for its time-slice. Such a procedure might incur a cost based on its being re-setup.

To recap therefore, the virtual manufacturing cell is an interesting concept which allows the flexible reconfiguration of shop floors in response to changing requirements. However, some major developments may have to be made in cell intelligence and in communications. Problems may also occur when workstations are reassigned. In the near future a compromise virtual manufacturing cell may arise, where reconfiguration will be carried out at relatively infrequent intervals (thereby overcoming some of the time-sharing problems) or where only a limited range of cell/workstation allocations will be possible (thereby reducing the communication problems).

HIERARCHICAL CONTROL SYSTEM EMULATOR

To aid the implementation of the AMRF a hierarchical control system emulator has also been developed. This emulator provides a detailed emulation of the individual control modules linked together in the AMRF control hierarchy (see Bloom *et al.*, 1984). This therefore allows, for example, the comprehensive testing of individual control modules prior to their implementation in the AMRF. It also allows the operation of an individual machine, or robot interactivity with the emulator, instead of with the rest of the AMRF. This, again, would aid the testing of control modules.

Within the AMRF control structure the control levels are

arranged in a hierarchy whereby each level takes commands from the level above and either processes commands for the level below or executes the command itself. The emulator has a separate module for each module of the control hierarchy; these emulator modules may reside on one or more processor. All transactions are carried out using state-tables and time-steps.

The state-tables include all the inputs, outputs, states and state transitions of each module (see Figure 4.4), so given particular inputs and state transitions, the outputs and state transitions can be readily deduced. Communication between modules is achieved by access to common memory 'mailboxes' with common input and output. Each mailbox is constructed so that only one module may write to it while several modules may read from it.

To run the emulator, the modules are combined into subsets which run as independent processors. This does require some synchronization which is accomplished by forced time-steps (or control cycles). At each time-step, the inputs and states are sampled and the outputs and state transitions are deduced. The user can decide how frequently each module should be invoked by using a scheduling interval; the module would only be invoked at each scheduling interval. In addition, the user can also specify the ratio of clock time to emulated time.

The emulator was originally written in the high-level language PRAXIS to run under the DEC VAX/VMS operating system, but a LISP based version is under development.

Comparison of AFMS and AMRF

Both the CAM-I advanced factory management system (AFMS) and the NBS automated manufacturing research facility (AMRF) include the use of a hierarchical framework for the planning and control of manufacturing systems, on the basis of a modular approach to the solution of complex problems. The two hierarchies have a rough equivalence in levels:

CAM-I AFMS	*NBS AMRF*
Factory	Facility
Job shop	Shop/cell

Manufacturing work centre	Workstation
Unit/resource	Equipment

The inclusion of the extra layer in the AMRF hierarchy is partly accounted for by the use of the concept of the virtual manufacturing cell; the inclusion of this concept leads to a somewhat divergent operation of the two structures. In the AMRF shop level, virtual cells are formed each with a cell controller. These cell controllers command several (possibly) workstations which each contain several machines. In the CAM-I AFMS there is no virtual manufacturing cell configuration and the job shop level controls all the manufacturing work centres.

Both the AFMS and the AMRF can support distributed data, so instead of data being stored in a centralized database, distributed databases are organized with some duplication of data.

To date, there has been no implementation of the AFMS. There does, however, exist a specification of data flows with data definition, but there is still a need for a verification of the data flows. By contrast, the AMRF architecture has been implemented with state-tables providing the logic source, although the virtual manufacturing cell configuration stage still awaits complete testing.

Overall, therefore, the CAM-I AFMS and the NBS AMRF present a somewhat similar control hierarchy, the major difference being the inclusion of the virtual manufacturing cell configuration stage in the AMRF. With the problems in implementation associated with the virtual manufacturing cell configuration, it may well be that this configuration stage implementation is postponed; the two structures would then appear more similar.

The following section deals with extracting from the two approaches a generic control structure suitable for a wide range of automated manufacturing systems.

GENERIC CONTROL STRUCTURE

Both approaches to the control of automated manufacturing systems have presented a hierarchical framework which decomposes commands from the higher levels to the lower levels. The two approaches have a large degree of similarity but they also, as discussed in the previous section, have

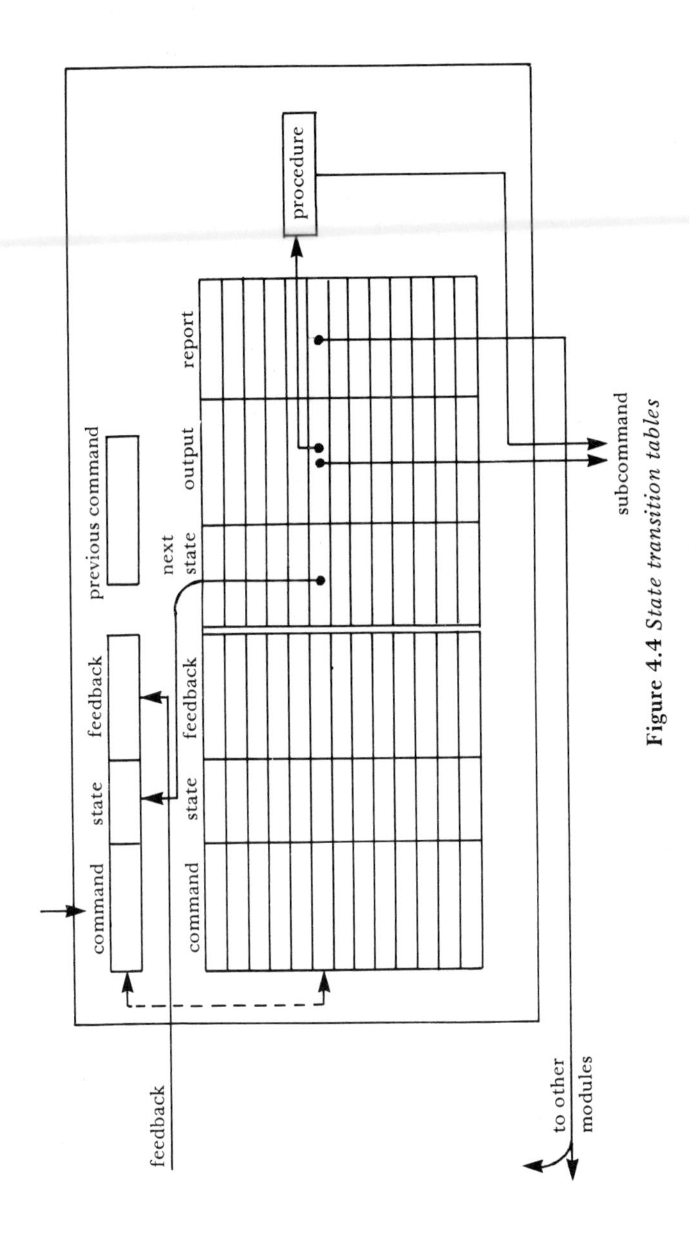

Figure 4.4 *State transition tables*

some differences. In this section, a generic control structure based partly on these two approaches is described, with the detailed description of each section being covered in later chapters.

The top level of the computer-based planning and control hierarchy for automated manufacturing systems is the corporate level, where activities such as computer aided design (CAD), process planning, MPS, and MRP are accomplished. At the lowest level of the hierarchy are the NC (or other) machines, robots, inspection devices, etc. The overall production planning and control operation for automated manufacturing systems is therefore concerned with taking requirements (and some data on, for example, process plans) from the highest level and controlling the lower levels to tie in as far as possible with these requirements. Therefore, a requirement for a particular end product probably passes through an MRP system to calculate the net requirements of parts; this net requirement, together with some of the process planning data, then goes to lower levels of the hierarchy. These levels then co-ordinate and control the machines, etc on the shop floor, in order to produce the net requirements at the correct time.

The number of intermediate levels of the hierarchy between the corporate or factory levels and the machine levels is two (job shop and work centre) for the AFMS and three (shop, cell and workstation) for the AMRF. The extra level for the AMRF is partly caused by the requirement for the virtual manufacturing cell reconfiguration; if it assumed that this is done relatively infrequently, then we could reduce the number of levels in the AMRF hierarchy to two. In any case, the number of levels is a somewhat artificial element. What is of more importance is what is achieved at each layer.

If we consider the major functions to be covered in taking the net requirements and decomposing these into commands for each machine, then probably two intermediate levels are sufficient. We shall term these shop and cell levels respectively. Including the top factory level and the bottom equipment level we therefore have four levels as follows:

1. *Factory level.* This level comprises the factory-wide computer systems

including MPS, MRP, CAD and/or process planning systems. These provide the driving input into the automated manufacturing system and this input is further decomposed at each lower level of the hierarchy.

2. *Shop level.* This level contains a shop control system (SCS), which is concerned with the overall co-ordination and control of a major section of the automated manufacturing system, comprising several cells. This level is therefore central to the production planning and control function. Interfaces are provided to each CCS and to the factory level software/hardware.
3. *Cell level.* This level has a cell control system (CCS), which is based on a micro- or minicomputer that controls one or more individual machines or other facilities (each probably with their own controllers). Each cell therefore consists of a group of one or more machines, robots, AGVs, etc. An interface is made both downwards to the equipment controller and upwards to the shop controller. The function of the CCS is to schedule and control the individual elements of the cell in order to achieve the commands or goals set by the shop level.
4. *Equipment level.* This level is concerned with the control of each individual machine or other facility. Usually the computer hardware at this level will take the form of a small computer or microprocessor to monitor and control the machine or other facility. The facilities at this level include NC machines, robots, automated storage and retrieval systems, AGVs and inspection equipment.

One major factor in choosing four levels is that this is usually the number of levels of computer hardware in the system. A schematic representation of these control levels is shown in Figure 4.5. Not all layers may be required for a particular automated manufacturing system. For example, where there are only a small number of cells that do not interact to any appreciable degree then the shop level need not be included.

Each level receives commands or goals from the level above and either carries out these commands or else further refines them to give commands for the level below. In this manner each level has *control* of the level immediately below it. Each level gives feedback on its status to the level immediately above it, so that the higher level can make informed decisions concerning the lower level. This command-feedback loop is illustrated schematically in Figure 4.6. A detailed description of the activities and function of each level is contained in later chapters. Suitable interfacing between the levels is of particular importance and recent

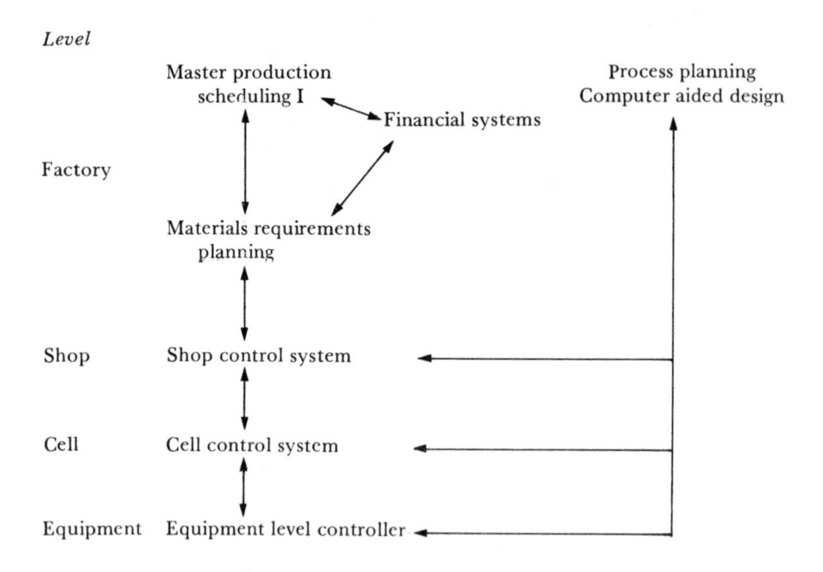

Figure 4.5 *Schematic representation of control levels*

developments in standardizing protocols (for example, General Motors' manufacturing automation protocol (MAP)) in automated manufacturing, can be of major benefit.

Conclusion

This chapter has described two major approaches to the hierarchical control of automated manufacturing systems, namely the advanced factory management system (AFMS) produced by Computer Aided Manufacturing Inc, and the automated manufacturing research facility (AMRF) produced by the National Bureau of Standards. Both these approaches break down the very complex problem of production planning and control into a number of layers so that commands flow down the hierarchy while feedback flows upwards. These two approaches provide the basis for a generic control structure, where output from the factory level systems is decomposed into commands for each machine (or other facility).

This generic control structure contains four levels; factory, shop, cell, and equipment levels, and each higher level has control over the level immediately below. The higher

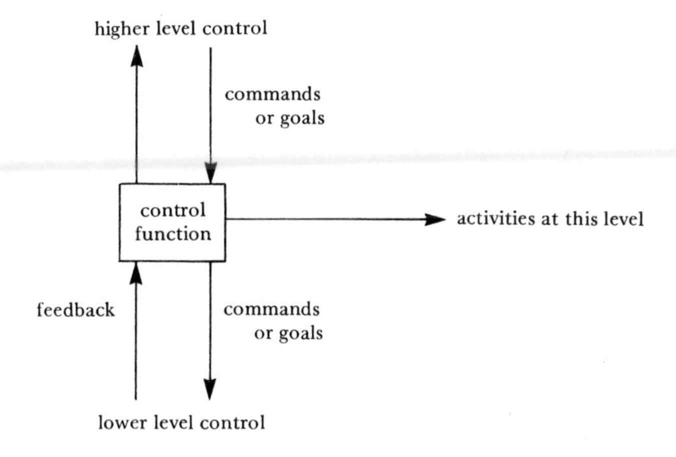

Figure 4.6 *Command-feedback loop*

level receives status feedback from the level below so that it has the knowledge necessary to make good decisions. More detailed descriptions of these levels are contained in Chapters 5 to 8.

CHAPTER 5

Factory Level Control

Introduction

This chapter briefly describes the structure and role of the factory level controller. The factory level forms the highest level of the control structure and deals with the relatively long-term strategy of the automated manufacturing system. It is, for example, at this level that decisions are made about the production rate and the due dates for jobs. Because decision making at this level involves taking a wide variety of factors into account, a human element in this process is desirable at least for the immediate future. At some future point, developments in areas such as artificial intelligence may mean that this human element can be reduced.

Aspects considered at this level include financial systems, computer aided design (CAD), process planning, master production scheduling (MPS), and materials requirements planning (MRP). The financial system deals with the financial analysis of the operation. CAD involves the design of the product. The process planning function plans the process operation of the job. MPS is concerned with the production of viable and effective schedules of end-products to be completed within a certain timespan (often a week) while the MRP process involves breaking down these schedules into requirements for parts and sub-assemblies. Time-phasing is also part of the MRP process, with the aim of ensuring that parts and sub-assemblies are available as and when needed.

These modules (financial systems, CAD, process planning, MPS and MRP) usually operate relatively independently. The present aim, however, is to link the computer systems involved more closely together, to form a more unified approach (this is sometimes termed computer integrated manufacture). Such integration efforts are currently in progress and should produce many benefits.

Although the control level is the factory level, it should be noted at the outset that this does not infer that the computer and other systems should necessarily be across the whole factory. It may be that the automated manufacturing system has its individual modules. For example, a significant trend in MRP has been to move away from large central processing facilities based on a mainframe computer, towards more local decentralized facilities based perhaps on a minicomputer. In this way the data transfer from shop floor to computer is eased and the MRP calculation can be done without the large infrastructure required for many large mainframe computers. It may well be that MRP, for example, is a local system to the automated manufacturing system and is not across the whole factory. The same could apply to the other modules at the factory level, where each may be specific to the automated manufacturing system.

Not all of the modules may be present for a particular facility. Where, for example, the production is essentially non-assembly then an MRP system may not be needed. Where the production is of standard products with little variation in the product range over time, then the CAD and process planning stage may not be needed. Each automated manufacturing system will be different, therefore, not just in the range of modules included at the factory level, but also in such aspects as the role of the human decision maker, the degree of integration between modules and in the number of modules that are specific and local to the automated manufacturing system.

Financial systems

Efficient monitoring of the financial performance of the automated manufacturing system is important in ensuring that the system maintains a good rate of return on capital employed. The role of the financial system is therefore to monitor the performance of the automated manufacturing system and to generate reports by translating the performance into financial figures. Aspects considered by the financial system include:

1. Inventory levels.
2. Work-in-progress levels.
3. Resources consumed (utilities, tooling, jigs/fixtures etc).
4. Payroll costs.
5. Depreciation.
6. Maintenance.
7. Insurance.
8. Overheads.

Typically, the inventory and work-in-progress costs will tend to be lower for automated manufacturing systems than for conventional manufacturing systems, since the manufacturing lead times are usually lower. However, costs such as depreciation, maintenance, insurance and overheads may well be higher, mainly due to the greater complexity and cost of the automated manufacturing systems. The overhead section may well include a contribution towards the CAD, process planning and other support functions; because of the greater emphasis on these functions in automated manufacturing systems, these costs may be higher than those for conventional manufacture. The use of this relatively expensive facility may well tend to be high – it is not unusual for machine use to be so extensive that replacements are required after five years, in comparison with the more common ten or fifteen years for conventional manufacturing.

The financial system uses data from a number of sources to carry out its analysis. Costs such as depreciation, maintenance, insurance, payroll cuts and overheads may be relatively fixed from one time period to another, and will require perhaps only a small amount of adjustment. The inventory levels, work-in-progress levels and resources consumed will probably vary markedly from time period to time period and some on-line data transfer may be desirable. The inventory and work-in-progress levels can be obtained directly from the MPS module, but the data required to evaluate fully the resources consumed may require a separate monitoring system.

Computer aided design

Where products have to be designed or redesigned at fairly frequent intervals, the design process becomes an important area for the automated manufacturing system. The major advantages of computerizing this activity are that it can result in a better quality design with a greater number of standard features, the productivity of design personnel can be increased, lead times can be reduced and the CAD process can help to generate the database necessary for manufacturing.

This creation of the database necessary for manufacturing is important in improving the overall CAD/process planning activities. CAD systems can generate the geometric data which can then be used in the process planning system, together with materials and machining specifications to produce the process plan. Much time can be saved by the direct use of the CAD output (by the process planning system accessing the CAD database) rather than by manually interpreting the CAD output drawings and manually inputting the geometric data to the process planning system.

Process planning

The process planning function develops detailed process plans both for the processing machines and for the material handling devices (for example, robots). The major portion of this process planning function will be concerned with three activities: first, to generate the manufacturing operation sequence; second, to produce the NC part program; and third, to produce the robot program.

The first activity has traditionally been a highly skilled task relying on the experience and judgement of the planner; but because they may apply slightly different criteria to the task, there may be a great variation between the operation sequences produced for the same part by different planners. A large amount of effort is being expended to encapsulate the skills of the process planner into a computer program or expert system in order to generate the process plans automatically. Many such approaches rely on the classification of parts into product families using group technology based approaches. This can simplify the decision making process but can involve an extra classification and coding stage. Inherent in many computer-based process planning systems is first, a machinability database to determine the proper speed and feed combination for the NC machines, and second, a computer-generated time standard to give the work content of tasks. This work content will, of course, tend to be less for automated manufacturing systems but it may include tool setting or other activities which remain a manual operation.

Master production scheduling I

The MPS function, briefly described in Chapter 3, is to determine a viable and effective schedule of production output for the time period under consideration (usually a week). Such an MPS is produced for every time period, stretching to the end of the planning horizon (often one year). In this manner production can be altered to tie in closely with expected customer demand and inventory status. The MPS function for automated manufacturing is essentially similar to that of conventional manufacture. However, there are some provisions which do slightly alter the MPS function:

1. Provision of a second MPS stage. The first MPS stage is similar to that already envisaged, and is likely to involve week-by-week production outputs. This output then, probably, passes through an MRP stage. The output from this goes into the shop control system where it is segmented into production for much shorter time periods. This activity is called master production scheduling II (MPS II) and is discussed in Chapter 6. The first MPS stage can be termed MPS I to distinguish it from MPS II.
2. Data input. The automated manufacturing system will have a large degree of computer control. The data used for MPS I, involving the status of the automated manufacturing system, is therefore likely to be both more accurate and more up to date since the data can be accessed relatively easily.
3. Capacity levels. Operation times tend to be well-defined within automated manufacturing systems. Consequently, the capacity level of the system can be known to a much greater degree of certainty than in conventional manufacture. Therefore, given both 2 and 3, MPS I is likely to be more accurate than in conventional manufacturing systems.

Materials requirements planning

If the production is assembly-oriented, in all probability the output from the MPS I function (which is a feasible end-product output) will be passed through an MRP system to determine the requirements for parts and sub-assemblies, and their timed release on to the shop floor. As with MPS

I, the function of the MRP system for automated manufacturing is essentially similar to that for conventional manufacturing (see Chapter 3). However, manufacturing lead times are likely to be shorter and better known for automated manufacturing systems and this may affect the MRP process, in that demands may need to be forecast over a much shorter horizon. The MRP system provides a list of jobs to be completed by the automated manufacturing system; this list is sometimes termed the 'work-to' list.

Data output to shop level

The data sent from the factory level to the shop level falls into (besides financial data) two major areas:

1. The 'work-to' list provided by the MRP system which gives the jobs to be completed and the due dates to be achieved. An urgency number may also be given to indicate the relative importance of the job.
2. The process plan provided by the process planning system which gives the routing for a job, the operation times, jigs/fixtures or special tooling required and perhaps the NC tape for machine or robot operation. It should be noted, however, that the full machine or robot program is not required at the shop level, but only the operation times for each job on each machine and the robot movement times. The full machine or robot program could be passed directly to either the cell controller, robot or machine thereby avoiding overloading the shop level with data transfer operations. This transfer to the cell controller, robot or machine could be done either by directly communicating or by using, for example, magnetic tape.

Conclusion

This chapter has briefly discussed the major blocks that form the factory level control. Included are financial systems, CAD, process planning, MPS I, and MRP. This factory level is the highest level in the control hierarchy and constitutes the level at which major external and managerial influences can be brought to bear. The time periods considered are relatively long in comparison with the lower levels of the hierarchy and detail is often only at the aggregate level.

The data output to the shop level mainly consists of a 'work-to' list and process planning information. However, much process planning information can pass directly to the cell, robot or machine.

CHAPTER 6

Shop Level Control

Introduction

This chapter describes the operation of control at the shop level. The shop control system (SCS) forms the second highest level of the hierarchy of control (as discussed in Chapter 4), and is mainly concerned with taking the requirements given by the factory level, and translating these into commands or goals for each cell. In doing so, it should take account of the constraints that apply both in terms of machine capacities and in terms of, for example, tooling or transport provision. In addition the SCS has to take account of factors such as the due date of particular jobs.

The SCS is central to effective production planning and control of the manufacturing system, in a manner which results in good *overall* system performance. 'Overall', because the aim is not to *sub-optimize* (ie to produce optimum solutions for each cell), but rather to produce a good performance from the system as a whole. It is for this reason that the SCS is so important. The SCS should have the data necessary to take a global view of the automated manufacturing system; it can then take the overall system performance as the criterion for evaluating decisions.

This chapter describes an approach to designing the SCS based on the operation being divided into two distinct processes, the first being master production scheduling II (MPS II) and the second being on-line scheduling. MPS II is concerned with organizing the 'work-to' list (the output from the factory level) into viable and effective subsets for completion in the automated manufacturing system in shorter time periods than those considered at the factory level. Whereas, for example, the 'work-to' list includes work to be completed within perhaps the next month, the MPS II function breaks this down into work to be completed in, perhaps, the next shift of eight hours.

The on-line scheduling function takes this requirement for work to be completed in, perhaps, the next few hours and produces a plan for loading individual cells in order, broadly to accomplish the requirements set by MPS II.

Taken together, MPS II and on-line scheduling therefore break down the requirements from the factory level into requirements for each cell. Both may need some computation time to do their task adequately; this time is usually available under normal or near normal operating conditions. Where there is a major change in the system's status, caused for example by a machine breakdown or other similar major change, then decisions may have to be made very quickly. The final section of this chapter describes a 'mailbox' approach where rules that can be quickly applied with reasonably good results are stored for fast access when needed.

Master production scheduling II

A possible production planning and control structure for the SCS is shown in Figure 6.1. Again, a hierarchy is used, this time to break the problem into two stages; MPS II and on-line scheduling. A hierarchical approach to the problem of production planning and control has been suggested by many workers including Bell and Bilalis (1982), Buzacott and Shanthikumar (1980), Eversheim and Fromm (1983), Fox (1982), and Stecke (1983).

The MPS II function level is concerned with extracting from the 'work-to' list (obtained from the factory level MRP system) a realistic portion of work to be done in a specified time period. The 'work-to' list may contain work to be done over perhaps the next month and the MPS II breaks that down into work to be done in a much shorter time period. The average operation times are important in deciding on the length of this shorter time period. With longer operation times, the time period considered by the MPS II function could be longer; where operation times are very short, the MPS II may only be concerned with time periods of perhaps an hour.

Superficially, the MPS II function is similar to the MPS I

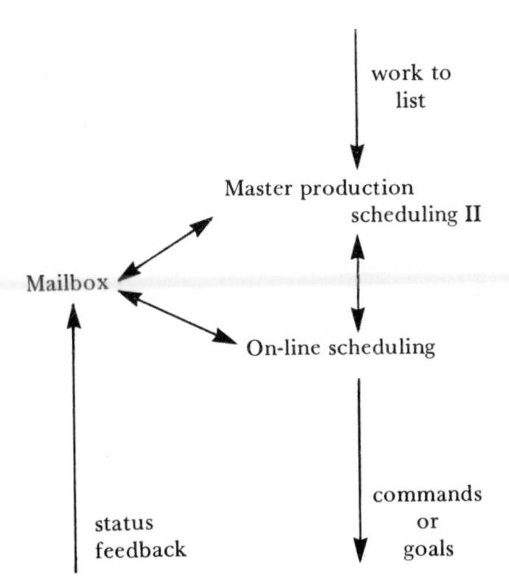

Figure 6.1 *Shop control system structure*

function that precedes the MRP calculations of net requirements (see Chapters 3 and 5). However, the two MPS functions differ in two significant respects:

1. The time periods considered are different – MPS I may be for weekly time periods whereas MPS II may be for time periods of a few hours.
2. MPS II may have to consider engineering details such as tools or jig/fixture supply.

Analytical methods of determining an efficient and viable MPS I have been discussed in Chapter 3; these approaches are capable, to a greater or lesser degree, of giving effective solutions for the MPS I function for conventional manufacturing. Obviously, any analytical solution will need to be treated with care to ensure that all important factors have been included in the derivation of the solution. Given this general warning, each of the above approaches has one or more drawbacks when applied to MPS II in a typical automated manufacturing system. The first drawback concerns the need to consider tooling and the provision of jigs/fixtures as a major constraint. Although the virtual capacity tool magazine has become technically feasible, it is the exception rather than the general case and a finite capacity tool machine is the norm at present. Carrie *et al.* (1983) cite the example of a major automated manufacturing system installation where a product mix of just 7 part types requires all 100 tool magazine slots at certain machines. In a practical automated manufacturing system there is also likely to be a limit on the number of particular jigs/fixtures available, because of the high cost of producing sophisticated jigs/fixtures, and there are therefore likely to be constraints on their use. This means that the MPS II function must not only take into account factors such as machine capacity as a constraint, but also other considerations such as the provision of tooling and jigs/fixtures.

The second drawback concerns the nature of the constraints themselves. Most approaches would consider the constraints as being 'hard', ie no overshoot of the constraint is possible. In many facilities this does not adequately reflect the decision making process; for example, the tooling required can exceed the machine capacity, providing that the time taken physically to change the tooling can be accommodated, presumably with some reduction in system performance. Any analytical procedure must therefore be capable of including constraints as both 'hard' and 'soft'.

The third drawback lies in the aims of the automated manufacturing system; often there are a wide range of partly conflicting goals (for example, high machine usage but low work-in-progress levels). There is also likely to be a great difference in aims from installation to installation; therefore the methods need to be able to handle this wide variation in goals.

The fourth and final drawback to many of the analytical approaches is that they often produce aggregate solutions which then have to be broken down to obtain the desired production rate for each product.

Some authors have considered the particular requirements of automated manufacturing systems: Stecke (1983), for example, has constructed a non-linear integer programming model for a particular automated manufacturing system and she was able to obtain some solutions using a standard mixed-integer programming system. This approach however, suffers from the usual problem associated with mixed-integer programming, namely that of high computation times. Queueing theory approaches have been considered by several researchers including Solberg (1976), Buzacott and Shanthikumar (1980), and Yao (1983). These queueing theory approaches can provide some useful results in the design phase of FMS but their usefulness in providing operational solutions has to be doubtful.

In the following section one approach, suggested by O'Grady and Menon (1985), is described. This approach, which is an adapted goal programming formulation, overcomes the drawbacks of other methods of dealing with the MPS II problem for automated manufacturing system.

AN ADAPTED GOAL PROGRAMMING APPROACH TO MPS II

The MPS II function is concerned with extracting from the 'work-to' list a viable work load for the automated manufacturing system over the next (short) time period. As indicated above, conventional approaches to MPS are often unsatisfactory when applied to the MPS II problem in automated manufacturing systems. Among other drawbacks, they usually deal with the problem only at an aggregate level, requiring disaggregation to obtain the MPS II. Furthermore, they usually lack a mechanism to consider detailed engineering requirements such as tooling constraints. O'Grady and Menon (1985) have described an adapted goal programming approach which overcomes the disadvantages of the conventional approaches.

The O'Grady and Menon (1985) approach is oriented to a general automated manufacturing system to include a wide variety of both 'hard' and 'soft' constraints. In addition, multiple goals are readily incorporated. The approach considers such aspects as: tooling requirements; linked groups of orders; machine capacity; alternative routes; due date considerations; volume of work-in-progress; and the expediting of certain orders.

O'Grady and Menon (1985) give a detailed account of how each of the above aspects can be formulated within their framework, the essential philosophy being that users can select, to their own particular requirements, individual modules as follows:

1. *Tooling requirements*. This module ensures that if a particular job is chosen, its associated specialized tooling is also allocated.

2. *Linked groups of orders.* This module allows the user to specify that all the orders in a particular group should be released simultaneously for production. This may be necessary where, for example, certain orders should be processed or inspected together.
3. *Tool magazine capacity.* Tooling magazine capacity may be limited; if this is the case this module allows the user to specify a 'soft' constraint on tool magazine capacity.
4. *Tool type availability.* Only a certain number of particular tool types may be available and this module ensures that this number is only exceeded on the basis of undergoing an expensive tooling acquisition procedure.
5. *Machine capacity.* This module allows considerations of capacity of the machines to be included.
6. *Alternative process routes.* This module ensures that only one route is selected for an order from a variety of routes that may be available.
7. *Due date consideration.* The consideration of delivery date or due date of an order can be of some weight in planning and controlling any automated manufacturing system. Orders that are nearing, or which have overrun, their due date will usually merit preferential treatment and this module allows for this.
8. *Product release level.* This module can be used to control the amount of products released in each planning period. This amount can be measured in volume terms or in any other terms (such as in monetary terms). In this manner this module can regulate the level of, for example, work-in-progress.
9. *Expediting certain orders.* This module can be used where it is deemed necessary to have a managerial override for trying to ensure that particular orders are included in the relevant plan. Both 'soft' and 'hard' formulation are possible: the 'soft' formulation allows the manager to specify a preference for a particular order to be selected, while a 'hard' formulation forces the order to be selected. Since the absolute forcing in the 'hard' formulation may produce difficulties elsewhere, the 'soft' formulation is usually preferred.

The manager wishing to use this approach of O'Grady and Menon (1985) would therefore choose from the above modules those which best fit the particular application. The constraints and attainment function are then formulated (Appendix I) and standard mathematical programming software is used to obtain suitable solutions. O'Grady and Menon (1986) describe the application of this approach to an automated manufacturing system in Scotland, containing six CNC horizontal boring machines each having a 100-pocket tool magazine. Use of the approach of O'Grady and Menon (1985) results in obtaining a plan that lies well within this tool magazine constraint and other constraints on the automated manufacturing system.

Computing time associated with the approach is not excessive: an average computing time for the system in Scotland was about 25 minutes CPU time on a VAX 11/780 minicomputer. Since this need only be run perhaps once a day, these times are entirely feasible for practical operations. O'Grady and Menon indicate that significant reductions in computation time can be achieved by, for example, only formulating as constraints those elements that are in practice likely to act as physical restraints. In the system in Scotland, for example, the tool magazine capacity was most likely to be a constraint on one particular machine and only this particular tool capacity needed to be placed in the module.

The output from the MPS II function is therefore a viable work load for the automated manufacturing system for a planning period usually measured in hours, not days. This work load plan will broadly achieve the goals of the automated manufacturing system while falling inside the inherent system constraints. Compromise may be necessary and the O'Grady and Menon (1985) framework provides a good vehicle for structured conflict resolution.

The computing times associated with this approach however mean that it is only suitable for use when sufficient time allows. At some future point, computing power may have risen to such an extent that the approach can be used in real time. For the foreseeable future though, the approach may have to be augmented by a 'mailbox' facility which contains some rules which can readily determine a reasonable MPS II without a signficant computing time delay. This 'mailbox' facility is discussed later in this chapter.

Whatever approach is used, the output of the MPS I function is a work load for the time period under consideration. This now passes to the second stage of the SCS, that of on-line scheduling.

On-line scheduling

The on-line scheduling function is concerned with taking the MPS II output (the viable work load for a particular planning period) and translating this into instructions to each cell control system (CCS), while taking into account both the status of the automated manufacturing system and the overall requirements. Again, we stress *overall*; we are concerned with the overall performance of the automated manufacturing system and not just with optimizing one particular cell, as this may give poor overall system performance.

The degree of detail required in the instruction to each CCS will depend on the ability of the CCS and the degree of decentralized control which is favoured. Where centralized control is required and/or there is limited decision making ability at the CCS level, the CCS requires relatively detailed instructions from the SCS. Conversely, where decentralized control is required *and* there is a reasonable decision making ability at the CCS level, the level of detail required is much less and broad goals for the cell to achieve may replace detailed instructions (this aspect is addressed more fully in Chapter 7). In this chapter, approaches which are capable of giving detailed instructions to the cells are described. These approaches can be 'de-tuned' to give only broad goals.

The background to on-line scheduling was discussed in Chapter 3. The major practical approach described was that of a simple fixed heuristic operating on the queue of jobs, whereby a priority is allocated to each job in the queue. The job with the highest priority is selected when the machine becomes vacant. The fixed heuristic approach has the advantage of ease of computation, but the disadvantage that the solution obtained may be poor when applied to a wide variety of system specifications. The requirements of the on-line scheduling function within an automated manufacturing system environment are such that good performance is needed across a wide range of operating conditions. In general, optimal approaches require large computing resources; a heuristic approach which adapts to a particular automated manufacturing system to give good results across a wide range of systems' operations seems to be the best compromise, both in terms of computing times and quality of the solution. Among the first to propose the use of adaptive heuristics were Fischer and Thompson (1963), with some later work done by Hershauer and Ebert (1975). The approaches used by these authors, however, suffer from the disadvantage that neither uses a unified format and consequently the methods have to be considerably altered for each application, making their use difficult in practice. The approach of O'Grady and Harrison (1985) is to use a heuristic that not only adapts to the manufacturing system but that is also expressed in a unified format; this approach is termed 'search sequencing'.

SEARCH SEQUENCING

The approach of O'Grady and Harrison (1985) is to use a priority rule operating on the queue of waiting jobs at each machine or process. Each job is given a priority index and when the machine or process becomes available, the job with the *lowest* value of priority index is given highest priority. The priority index, P_i, is given by:

$$P_i = A_i T_i + B_i S_i \quad (1)$$

where P_i is the priority index for job i at its current stage; A_i is a $1 \times m$ coefficient vector for job i; T_i is an $m \times 1$ vector which contains the remaining op times for job i in process order; B_i is the due date priority coefficient for job i; S_i is the due date slack for job i (defined as [due date for job] – [current value of date]); m is maximum number of processing stages for jobs 1 to i.

Equation (1) represents the general case where each job i has distinct values of A_i and B_i. Within the format of Equation (1) a wide variety of the conventional fixed heuristics can be expressed with A and B as follows:

1. *Shortest imminent operation time.*
   (Rochette and Sadowski, 1976)
   $A = (1, 0, 0, 0, \ldots.., 0)$
   $B = 0$
2. *Slack sequencing.*
   (Rochette and Sadowski, 1976)
   $A = (-1, -1, -1, -1, \ldots.., -1)$
   $B = 1$
3. *Due date sequencing.*
   (Cheng, 1982; Panwalker *et al.*, 1982)
   $A = (0, 0, 0, 0, \ldots.., 0)$
   $B = 1$
4. *Longest remaining processing time.*
   (Fischer and Thompson, 1963)
   $A = (-1, -1, -1, -1, \ldots.., -1)$
   $B = 0$

Equation (1) represents a general format for expressing the priority of a job. The equation also provides the ideal basis for making the priority index P_i adapt to the manufacturing system. This can be achieved by the following:

1. Give initial values of A_i and B_i.
2. Simulate the behaviour of the manufacturing system to evaluate the overall performance with the particular values of A_i and B_i.
3. Use a search routine or other method, to give another value of A_i and B_i and then repeat step 2. Stop the process if steps 2 and 3 have been repeated several times and little improvement in performance is detected. The values of A_i and B_i so determined can then be used to sequence the jobs.

The search technique used by O'Grady and Harrison (1985) is a modified version of the Hooke-Jeeves (1961) pattern search routine. The modifications are made to reduce the risk of the search becoming trapped in local minima. Other search techniques could be used, however, and their efficiency may well suit particular automated manufacturing system environments.

Simulation is an integral part of the search process and the manufacturing system may have to be simulated over its planning horizon several times before a satisfactory solution is obtained. Computing times associated with the simulation therefore become crucial. It may well be found that an approximate simulation model can give an adequate representation of the behaviour of the manufacturing process. As developments take place in computing hardware and software the need to use such approximate simulation models will lessen.

Equation (1) is a general representation of priority index and, as such, contains a large number of variables. For a 7-stage, 60-job system there would be up to 480 independent variables. The computation required to evaluate such a number of variables would be considerable. To reduce the computation a number of simplifications can be made to A_i and B_i as detailed by O'Grady and Harrison (1985):

1. Allocate the same value of A_i to all jobs i. An alternative would be to divide the jobs into a number of groups based on some criterion such as product families and then to have the same value of A_i for each group.
2. Ignore later stages and operations in the $A_i \times T_i$ term, thereby reducing the size of the vector A_i.

3. Reduce the number of different elements within A_i so that A_i instead of being a 1 × m coefficient vector with m different elements is now still a 1 × m coefficient vector but with fewer different elements. For example, where only four different elements are required then a possible format is $A_i = (a_{1i}, a_{2i}, a_{3i}, a_{4i}, \ldots, a_{4i})$.
4. Reduce the number of different values of B_i. In the most drastic simplification, the same value of B_i could be allocated to all jobs. Alternatively as in 1 above, the jobs could be divided into a number of groups each with its own value of B_i.

Each of these simplifications has to be considered in turn for each particular automated manufacturing system and a suitable combination of simplifications 1-4 above selected.

The on-line scheduling function, therefore, by the use of adaptive heuristics as described above, produces a detailed schedule of activities to be passed to the CCSs for action. Feedback from each CCS updates the simulation model used in the on-line scheduler.

Specific data requirements

The functions that are specific to automated manufacturing systems (ie MPS II and on-line scheduling) both have their own particular data requirements and these requirements are briefly described in this section.

The MPS II data requirements depend on which aspects are included. The basic data input is the 'work-to' list containing the job number and quantity required. The other data required depends on which modules are selected, with the following as possibilities:

Tooling requirements.
Linked orders.
Tool magazine capacity.
Tool type availability.
Machine capacity.
Alternative routeing
Due dates.
Product release levels.
Expedited orders.
Preference weights.

For the on-line scheduling stage, the data requirements are:

Viable work load (from MPS II stage).
Operation times.
Due dates.
Performance measure (for example, cost, throughput times or lateness).

Both the MPS II and on-line scheduling stages rely on a reasonably good model of the automated manufacturing system. Data feedback from the system is therefore an important factor in maintaining a good model of its present status. The data necessary for this includes the status of:

Each job.
Each machine/process.
Each transport or inspection device.
Tooling and/or jigs/fixtures.

Such data flows can be incorporated into the control structure. Data on the status of each cell is fed back from the CCS to the SCS which then handles the MPS II and on-line scheduling functions.

Mailbox approaches

The SCS has been broken down into two levels: MPS II and on-line scheduling. These two functions take as data input the 'work-to' list from the factory level, and also some process planning data and feedback on the status of the cells. Using these data inputs the two functions produce commands or goals to be passed to the CCSs. So that as the MPS I alters, thereby producing (via the material requirements planning system) a different 'work-to' list, the SCS recalculates the commands or goals to be sent to the CCS. Since changes to the MPS I or to the 'work-to' list occur relatively infrequently, the computing times associated with the methods discussed for these two functions are not particularly crucial. However, there is another cycle of events in which the computing times begin to become important. This cycle of events is associated with some major unforeseen occurrences that demand a relatively quick decision to be made. For example, a mechanical breakdown would require a number of decisions to be made, including the re-routeing of jobs, the provision of tooling or other machines and the organization of transport for these re-routed jobs. The approaches so far discussed require, at present, a reasonable amount of computing time and so are relatively unsuitable for the task of quick decision making. It may well be that at some future point the algorithms, etc will have been suitably refined and/or computing hardware power will have been increased to allow this real-time decision making to be done using these approaches. For the present though, another way of progressing is desired, which both gives reasonable decisions and has a real-time response.

An approach which is capable of giving a fast response is that of using 'mailboxes' which contain a number of rules in a mailbox matrix to be used when desired. The particular slot in a 'mailbox' array is accessed when certain conditions are satisfied. Such an approach is similar in concept to the state-table approach adopted by the AMRF. Under the mailbox approach therefore, a number of rules are arranged so that they can be readily accessed. These rules may vary from fixed specific rules, to heuristics in an IF-THEN format. For example:

Mailbox Slot 1
IF machine C is not useable THEN re-route all jobs requiring machine C to machine F

Mailbox Slot 2
IF jobs are to be sequenced on machine F THEN use priority rule a with coefficients x and y

The manner of operation of the mailbox array is then similar to a production rule approach of artifical intelligence. The mailboxes are examined to see which rules apply. The rules that apply may then be ranked as to the sequence in which they are applied and they are then applied in that sequence. Some points about a mailbox approach should be noted:

1. The number of rules should be limited – too many rules may mean that too much time will be associated with accessing the rules, deciding which rules apply, determining the sequence (if there is more than one rule applicable), and then applying these rules.
2. Conflict resolutions – if more than one rule is triggered then there is a possibility that the rules may conflict. Some system for allocating priority to the rules may help to resolve the conflict.
3. The rules may link with the MPS II and on-line scheduling approaches already discussed. For example, the coefficients x and y in the mailbox slot 2 above may be set initially by the user, but could be refined gradually by the on-line scheduling approach discussed in this chapter, in order gradually to improve the performance of the rule.

This mailbox approach is the basis of the production

logistics and timings organiser (PLATO) which uses artificial intelligence techniques. PLATO is currently under development at North Carolina State University.

Conclusion

This chapter has discussed the shop control system (SCS) which takes commands consisting of a 'work-to' list from the factory level, with other data from the process planning system and the CCSs. From these inputs the SCS produces commands or goals to be passed to the CCSs. The degree of detail required in the commands or goals will vary with the particular automated manufacturing system installation. Where, for example, a centralized control structure is desired and/or the CCS has limited decision making ability, the commands passed to the CCS will probably be fairly detailed. If a decentralized control structure is desired and the CCS has good decision making abilities, the commands may be only in the form of a broad goal. This aspect is discussed further in Chapter 7.

The operation of the SCS can be broken down into two levels: MPS II and on-line scheduling. An adapted goal programming approach to provide good solutions to the MPS II problem has been described; the advantages of this approach being that it can consider a wide variety of aspects including machine, tooling and jig/fixture supply. In addition, it can accept those constraints that are fixed as well as those that are capable of movement, perhaps at some penalty.

The main approach presented for on-line scheduling has been adaptive heuristics where the heuristic rule adapts to a particular manufacturing system. The degree of detail within the rule's operation can be set by the user, so it can be used for both highly detailed commands or for more broad goals to be sent to the CCSs.

Both of these approaches need a reasonable amount of computing time in order to work properly; this time may not be available where there is a major unforeseen event, for example, a machine breakdown. The two approaches can therefore be augmented by a rule-based approach contained in what are termed mailboxes. These rules are arranged in an IF-THEN format. When problems occur, the rules in the mailboxes can be examined and those that are triggered are then arranged in a sequence for application. The rules themselves may be derived from other approaches (the adapted goal programming approach and the adaptive heuristic approach) and would then provide a strong link between the elements of the SCS. It could also lead to a gradual improvement in the quality of the rules.

CHAPTER 7

Cell Level Control

Introduction

The previous chapters have given a description of the planning and control hierarchy (Chapter 4), a background to the factory level (Chapter 5) and the shop level (Chapter 6). This chapter describes cell level control, which is beneath the shop level in the hierarchy of planning and control.

The cell control system (CCS) has to be designed to take the commands or goals from the shop control system (SCS) and translate these into specific commands or goals for individual items of machinery or other items at the equipment level. As discussed in Chapter 6, the degree of detail required in the commands or goals from the SCS will depend on the circumstances: where a centralized control structure is desired and/or the CCS has limited decision making ability then the commands from the SCS will have to be fairly specific. Conversely, where a relatively decentralized control structure is desired *and* where the CCS has good decision making abilities then only broad goals will need to be set by the SCS for the CCSs to carry out. The capabilities of the CCS are therefore important in determining first, the type of control that is possible and second, determining from this the command or goal data flows between the SCS and the various CCSs. Categorizing the CCS decision making abilities is discussed in the first section of this chapter.

Having categorized the various CCSs, the composition of a cell is another important aspect then to be considered, and is discussed in the section entitled 'What is a cell?'. A cell may vary from a single machine to several machines with local storage and material handling and a general CCS architecture must be capable of dealing with a wide variety of cell configurations.

The background to the operation of the CCS is then discussed by describing the modes in which the CCS can operate. Four distinct mode types are described, ranging from a highly centralized mode with direct detailed commands from the SCS, to a highly decentralized mode with only broad goals being set by the SCS. In practice, particular implementations of the CCS may well fall between two particular modes. CCS instruction or processing requirements via commands or goals are received from the SCS, and the function of the CCS is to ensure that these instructions are carried out as far as possible. The degree to which the CCS is capable of operating independently of the SCS will depend on its own decision making abilities and on its local memory. A CCS with a high decision making ability and a large local memory can be capable of a high degree of independent operation; it can make many decisions and access local data storage of, for example, NC programs without recourse to the SCS.

The storage of the main programs for manufacture including both robot and NC part programs can take place in a number of areas:

1. *Factory level.* Robot and NC part programs can be considered to be developed and then stored at this level, until required to be transmitted to the lower levels (or they can sometimes be transmitted prior to actually being required).
2. *Shop level.* Some robot and NC part program data is required at this level, but it is unlikely that the full programs will be stored solely at this level. Instead they will probably be transmitted directly to lower levels. When there is some limitation on storage at these lower levels, storage may well take place at shop level.
3. *Cell level.* Robot and NC programs could be stored here after being transmitted from the factory or shop levels, and then sent to the machines and robots as required. It is especially useful to store programs at this level where, for example, there is only limited storage available at the equipment level.
4. *Equipment level.* For ease of access, it is perhaps best if the full programs are kept at this level. However, some provision has to be made for the essential portions of data necessary for shop and cell level planning to be made available to these levels. In addition, any engineering changes made at the factory level must be transmitted to this level.

CCS classification

There are essentially two factors at cell level which influence the data and control structures:

1. *Decision making ability*. These CCSs with high decision making ability need only broad goals to be set.
2. *Local memory access in the cell*. This memory may be at the cell level or the equipment level. A reasonably large memory is required to store the NC and robot part programs that are produced at factory level. A low memory within the cell would mean that part programs would have to be transmitted at regular intervals from outside the cell.

If these two factors are combined, there are four categories of CCS with intelligence (taken to mean the ability to make decisions) and with memory (taken to mean the memory *within a cell* that is at both cell and equipment levels):

Class I CCS. High memory, high intelligence devices. This class of CCS is capable of both local storage of large data files and of some local decision making.
Class II CCS. Low memory, high intelligence devices. This class of CCS is capable of some local decision making but must receive, either from the shop or factory level or perhaps from a Class III or Class I CCS, part programs, for example.
Class III CCS. High memory, low intelligence devices. This class of CCS is capable of local storage of, for example, NC part programs, but still requires detailed instructions from the shop level.
Class IV CCS. Low memory, low intelligence devices. This class of CCS has to receive both detailed instructions and perhaps NC tape data from external sources (usually the shop and/or factory level).

Therefore, for both Class IV and Class III CCS, detailed instructions for actions to be taken within the cell have to be passed from the shop level. The role of the CCS in this case is that of storing a limited set of instructions to be followed. Whereas for both Class II and Class I CCS the instructions from the SCS need only be overall instructions or goals, and the function of the CCS is to organize the cell activities to broadly tie in with these goals.

For both Class IV and Class II CCSs, large data files have to be regularly transmitted to the cell either by direct computer links or by physical movement of, for example, magnetic tape. For Class III and Class I CCSs, there is a reasonably large amount of memory available within the cell and the only times that new part programs need to be transmitted to the cell or equipment levels are when there is a change to an existing part program, or a new product or part program.

What is a cell?

The sections above have briefly described the location for storage of NC and robot part programs, and have identified a classification scheme for the various CCSs. In this section we take a step back to look at the various configurations that we could find within a CCS.

The actual composition of a cell can vary tremendously but all cells will usually be made up of:

1. CCS computer processor.
2. NC machines or other processing equipment.
3. Robots or other material handling devices.
4. Storage elements.

A further description of each of these constituents is contained in Chapter 8. A cell will consist of some of these constituents linked together through the CCS. The only essential ingredient is the CCS computer processor, the other constituents being added to achieve the desired format. A cell that is primarily concerned with machining would contain one or more NC machines, robots or other material handling devices as well as perhaps an input buffer storage area and/or an output buffer storage area. A typical cell layout is shown in Figure 7.1. Another cell might be primarily concerned with assembly and would usually contain assembly machinery, including robots, instead of NC machines. Yet another CCS might be concerned with transport between cells, and would therefore be primarily involved in controlling transport or other material handling facilities such as automated guided vehicles or conveyors.

The size of a cell can vary markedly. A small cell is shown in Figure 7.1; other cells could be larger than this. Probably the largest cell configurations have been those proposed by the National Bureau of Standards in their automated manufacturing research facility (AMRF) where the use of virtual manufacturing cells has been proposed (Chapter 4). Under this proposal, the whole shop is sub-

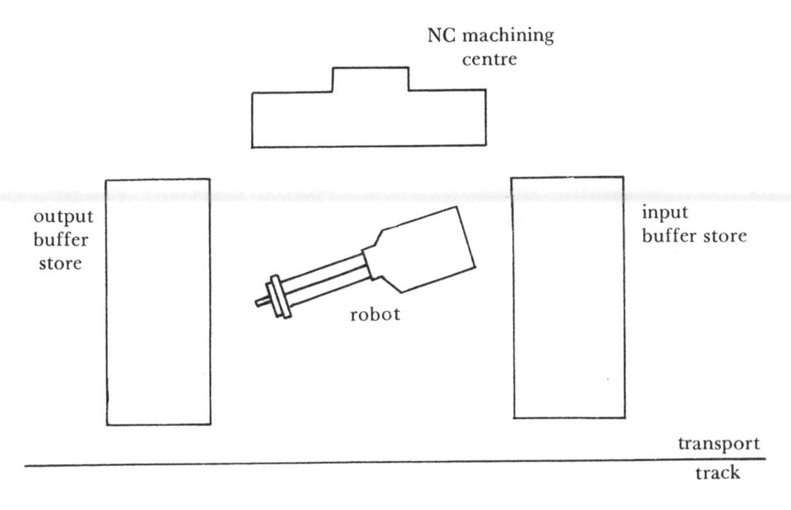

Figure 7.1 *Typical cell layout*

divided into virtual manufacturing cells in a dynamic manner (ie at relatively frequent time intervals) on the basis of a group technology classification. Jobs are grouped by this classification into product families of jobs with relatively similar processing requirements. The machines and other facilities on the shop floor are then arranged into virtual manufacturing cells to manufacture each product family. This is not a physical rearrangement, but rather one that involves a reallocation of machines to different CCSs. Both the process of grouping jobs into product families and of allocating machines to virtual manufacturing cells is done at relatively frequent time periods. The result of all of these procedures is that each virtual manufacturing cell is capable of carrying out all the machining or other requirements for a product family. As such they would tend to be both large and in a flow line arrangement, where jobs in a particular family having finished on one machine would all go to the next machine. This flow line type of operation would probably simplify the control problems. However, some other aspects of virtual manufacturing cells (discussed in Chapter 4) may well complicate control. These other aspects include the dynamic reallocation of machines to cells and the possibility of 'time-sharing' machines between cells. It may be some time before the tools necessary to deal with these aspects are fully developed; because of this the remainder of this chapter deals with more conventional approaches to cell architecture although, of course, the concepts inherent in virtual manufacturing cells should be borne in mind.

CCS operational modes

The CCS can be designed to operate in a number of modes, depending on the structure, configuration and requirements of the automated manufacturing system. These include four major modes:

1. Highly centralized mode.
2. Loading mode.
3. Itemized mode.
4. Decentralized mode.

This list is not exhaustive – there are other possible modes but they will tend to be subsets and/or combinations of the four major modes above.

HIGHLY CENTRALIZED MODE

Under this mode of operation, decisions are made, on the whole, at shop level by the SCS. These decisions are transmitted to the CCS to be carried out. This means that little or no decision making is done at the cell level. The communication channel between the SCS and CCS is vital to the operation of the cell; if this communication channel is broken for any reason the cell activity ceases. It also means that if, for example, the SCS processor is overloaded at any point, with a long response time, then the cells could be kept idle until the SCS is able to respond. The level of communication between the SCS and the CCS can be illustrated by the following dialogue, which describes the information passed between the SCS and the CCS.

Source of dialogue	*Information passed*
CCS	Transport device 12 arrived at input station 2.
SCS	Load transfer device 7 with program XTR12AM362.
CCS	Transfer device 7 loaded O.K.
SCS	EXECUTE program on transfer device 7. (This moves part from transport device 12 to machine 16)

CCS	Starting execution of transfer device 7.
CCS	Execution completed O.K.
	(now machine 16 is loaded)
SCS	Load machine 16 with program XMC16PL214.
CCS	Machine 16 loaded O.K.
SCS	Execute program on machine 16.
CCS	Executing program on machine 16 (now the part is being machined).
CCS	Execution on machine 16 completed O.K.
	etc.

The above dialogue is not the actual control communication which takes place, but it serves to indicate the level of control exercised by the SCS over the CCS. Therefore, the control structure in this case causes the SCS to issue instructions to the CCS which then performs them as required. The CCS does not issue primary instructions itself.

Feedback is provided by the CCS on any event which takes place; this would include confirmation of requests, signals indicating start and end of processes and any error conditions occurring. Error conditions are signalled by intelligent resources and sensors which monitor other resources; these error conditions are then passed to the CCS. For example, the program may cause a robot to move a pallet to a certain location and if sensors indicate that this has not occurred, the cell generates an error condition. In practice, this would be essential in some form and therefore must be catered for in the design of a CCS.

Decisions are made by the SCS on the basis of comprehensive feedback from the CCSs; this feedback gives the SCS an accurate view of cell level operations and enables it to investigate decision making constructs with a microscopic view of cellular activities.

The major advantage of the centralized mode is that small automated manufacturing systems can be commissioned fairly rapidly: the control resides in the SCS and the CCSs follow the commands made by the SCS. Centralization can ease the programming problems. It is for this reason that many of the first-generation automated manufacturing systems have followed the centralized mode. A possible problem is the dependence on the SCS: if the CCS/SCS link were to be broken then cell activity would cease; if the SCS were to be overloaded then cell activity would slow down.

LOADING MODE

Using the loading mode, a cell is *loaded* with a quantity of work to be done by that cell during a specified time period. We can call the quantity a 'work lump'. At the end of that time period, it is expected that another work-lump will be loaded. The CCS function is to schedule and control this work-lump in order to complete all work within the given time period.

Decisions on possible cell loads are made by the SCS on the basis of feedback from the cells. The SCS must therefore have an accurate and up-to-date status report on each cell in order to make good decisions, and the feedback from each cell is important. Control is less centralized than with the previous highly centralized mode; if the communication channel between the CCS and SCS were to be broken for any reason, then the CCS could still carry out the activites on its work-lump but it would be forced to stop cell activities when this work-lump was completed.

The sequence of activities within the CCS are:

1. Receives work-lump notification from SCS together with the time that it is due. This work-lump comprises work arriving from (perhaps) several other cells.
2. The CCS evaluates the resources (machines, tools, jigs/fixtures etc) required to complete the work-lump in the specified time. It is at this point, before the work-lump physically arrives, that the CCS notifies the SCS if there will be any problems in completing the work-lump in that time. If there are problems, the SCS has the option of either letting the cell proceed with the activities even though the time period may be overrun, or of calculating a new work-lump for that cell (in which case steps 1 and 2 are repeated).
3. The work-lump physically arrives. Where the work has come from a single previous stage (another cell or a warehouse), it may arrive as a single entity. Where it has come from a variety of previous stages, it may arrive over a short period of time.
4. The CCS then controls activities in the cell to carry out the detailed schedule evaluated in step 2.
5. When the work-lump is completed (or nearly completed) the CCS notifies the SCS and transport is arranged.

The advantages of the loading mode are first, its relative simplicity and second, its greater independence over the centralized mode from direct links to the SCS. However, the loading mode may have some disadvantages: first, there could be problems in ensuring that the work-lump arrives more or less at the same time; and second, the provision of a work-lump at each cell would tend to raise the work in progress levels. Taking a simple example of this second point; suppose that the average operation time for a job at a cell is x minutes and the average number of jobs in a work-lump is n, then:

$$\text{Average work-lump} = n \times \text{minutes}$$

This means that each job in that work-lump would be nx minutes at that cell, even though its actual operation time might be considerably less than nx minutes.

ITEMIZED MODE

Under this mode of operation, the SCS loads the cell with individual jobs at relatively frequent time intervals so that when a job finishes at one cell it can immediately be transported to the next cell to join the queue of work there. The cell therefore receives frequent additions to its input queue. Therefore, this mode differs from the loading mode in that the work-lump is a single job, and when a job is finished in a cell it does not have to wait until all the other jobs in the work-lump have been completed.

The decision as to which cell to send a job to is made by the SCS and data on the jobs allocated to each cell are sent directly from the SCS to the allocated cell. Again, feedback from the CCS to the SCS is important in achieving good decision making at the shop level. In the event of some break in communication between the CCS and the SCS, activities can still continue with the cell until the jobs in the input buffer store are exhausted, provided of course that there is sufficient storage space within the cell for the completed jobs, or that transport can be organized without communication to/from the SCS.

The sequence of activities within the cell are rather similar to the activities of the loading mode, as follows:

1. The CCS receives notification from the SCS that a job will arrive, with details about the job (for example, estimated time of arrival, due date, operations required, etc).
2. The CCS produces a detailed schedule of activities to be completed

within the cell to achieve the requirements sent by the SCS (see the section on 'loading mode' for further details).
3. The job physically arrives.
4. The CCS carries out the detailed schedule determined in step 2.
5. The CCS notifies the SCS when a job is completed or nearly completed.

Note that because jobs arrive singly instead of as a work-lump, the CCS has to complete the detailed schedule of cell activities (step 2 above) more frequently than in the loading mode; this may mean that computing times associated with the detailed scheduling become an important factor.

The advantages of the itemized mode are first, that a certain degree of decentralization is inherent, so the cell *can* carry on with activities in the event of a SCS-CCS communication break (this is primarily an advantage over the highly centralized mode). Second, lead times and work-in-progress levels will tend to be lower than in the loading mode, since jobs will not have to wait until the work-lump is completed as they would in the loading mode.

One possible disadvantage with this itemized mode is that there is still considerable SCS-CCS data communication and this may overload the SCS or the communication highways; if this is the case, the final mode discussed may be preferable.

DECENTRALIZED MODE

In this mode there is minimum communication between the SCS and the CCSs. The SCS's primary function is to start each job off at its first cell and pass enough information to this first CCS so that the CCS can arrange for the job to be transported to its second cell. The first CCS also has to arrange for enough information to be passed to the subsequent cells to permit the completion of the job.

The communication process takes the following form:

SCS to CCS of cell 7 (the job's first cell)	Job ABC123, part number XX3YZ; lotsize 20; estimated arrival time 12.32; routing cell numbers 7, 2, 6, 11, 1; operations at each cell: cell 7, operation number 373.2, robot program A21, estimated finish time 17.22; cell 2, operation number 118.7, robot program D39, estimated arrival time at cell 2, 17.39, estimated finish time at cell 2, 17.58, cell 6 etc.

The first cell (ie cell 7) then evaluates the resources and schedules activities in order to complete all jobs by, or near to, their allocated estimated finish time. If this cannot be done, the CCS notifies the SCS and the SCS has the option of either agreeing to a revised finished time or of reloading the cell and starting the process again. Assuming that the allocation by the SCS is agreeable, the CCS notifies the SCS.

CCS (cell 7) (to SCS)	O.K.

The CCS then carries out the detailed schedule of activities including the processing of job ABC123. When this job nears completion, the CCS notifies both the transport CCS and the job's next CCS (cell 2):

CCS (cell 7) to transport CCS	Job ABC123, part no XX3YZ; lotsize 20; estimated finish time 17.21; transport to cell number 2, estimated arrival time 17.39.
CCS (cell 7) to CCS (cell 2)	Job ABC123, part number XX3YZ; lotsize 20; estimated arrival time 17.39; routing 2, 6, 11, 1. Operations at each cell: cell 2, operation number 118.7, robot program D39, estimated finish time at cell 2, 17.55; cell 6, operation number 971.6 etc.

The transport CCS then schedules the transport in order to carry out the requirements specified by CCS (cell 7); if this is not possible, it notifies the CCS (cell 7) which then re-evaluates its requirements. If there are unsurmountable difficulties, the CCS (cell 7) notifies the SCS which then reloads the shop.

Meanwhile, CCS (cell 2) reschedules activities within the cell to accommodate job ABC123 (note that as job ABC123 passes through the shop, the data transmitted between cells becomes less). Again, if CCS (cell 2) is unable to complete all jobs by, or near to, their allocated estimated finish time then it notifies the SCS and the SCS reloads the shop.

In this manner the SCS is managing by exception. It is only when there is a significant departure from planned activities that the SCS is involved; otherwise the shop carries on with the allocated loading. The frequency with which the SCS is involved will depend on how tightly the finish times are defined; with slack time built into the finish times it becomes more likely that the CCS can complete the job by the given finish time. This slack may mean, however, that work-in-progress and lead times are longer and a compromise slack may have to be used.

The advantages of the decentralized mode are that it can be designed so that the shop more or less runs itself, with the SCS only being involved to start a job or when major problems arise. In this way the SCS-CCS link is less vital and if there were a break in this communication channel the shop might well be able to carry on activities for some time.

A possible problem in implementation could be the necessary CCS-CCS network carrying data from CCS to CCS; care needs to exercised, using a network that does not easily become overloaded or corrupted by the electrical noise usually associated with shop floor activities.

Conclusion

This chapter has provided an overview of cell level control which functions to take requirements from the shop level and to translate these requirements into specific cell activities. Four classes of CCS have been defined: Class I CCS being high memory, high intelligence devices; Class II CCS being low memory, high intelligence devices; Class III CCS being high memory, low intelligence devices; and Class IV CCS being low memory, low intelligence devices. The memory factor will determine where the NC and robot programs are kept, while the intelligence factor will determine the amount of decision making that can be achieved within the CCS.

The second aspect considered in this chapter has been the nature of a cell, and the typical constituents of a cell were discussed. Most cells will tend to be small processing cells containing, for example, an NC machine, a robot, an input buffer store and an output buffer store. The largest cell configurations proposed have been by the AMRF who suggest the use of virtual manufacturing cells containing all the processing equipment necessary to completely process a group technology product family.

The various operational modes for the CCS have been detailed. These have been categorized into four modes: highly centralized; loading; itemized; and decentralized.

The highly centralized mode depends on the SCS issuing specific commands to each CCS on each activity that takes place within the cell, so that all decisions are made at the shop level and the role of CCS is simply to direct these commands to particular machinery. In the loading mode the SCS loads each CCS with a work-lump that is a portion of work to be done in a specified time interval. The CCS then has to arrange activities within the cell in order to complete the work within the timespan given, and therefore greater CCS decision making capabilities are required in this mode than in the highly centralized mode.

The itemized mode is similar in many respects to the loading mode, the difference being that the jobs are loaded individually to each cell instead of being grouped together as a work-lump, as in the loading mode.

The decentralized mode is an adaptation of the itemized mode, in that the SCS only starts a job at its first cell. The CCSs then take over and communicate with each other to progress the job through the shop. The SCS is only involved if major problems occur.

The advantages and disadvantages of each of these approaches have been given and each will have its own range of automated manufacturing systems for which it is most suitable.

CHAPTER 8
Equipment Level Control

Introduction

The equipment level is the lowest level of the production planning and control hierarchy. Above the equipment level sit the factory, shop and cell levels and as we descend the hierarchy broad objectives specified in the master production schedule I (MPS I) have become more and more detailed commands. The timespan encompassed at the higher levels has also gradually decreased, so that the time bucket of perhaps a week at the MPS I level, has been reduced to a matter of minutes or seconds at the cell and equipment levels.

What acts as the main data entry from the higher levels to the equipment level is the output from the cell control system (CCS). This output from the CCS consists of specific commands to individual items of equipment, the CCS acting as a co-ordinator of cell activities. The equipment level controllers then have to interpret and obey these commands. The communication between the CCS and the equipment level controller seems superficially to be straightforward, however this communication and equipment co-ordination is one of the major problem areas in automated manufacturing systems. The problem is likely to ease in the near future, as general protocols such as General Motors' manufacturing automation protocol (MAP) come into use.

This chapter is not meant to provide a detailed background to automated equipment, but to provide a brief overview of the function and structure of the equipment level controllers. The chapter first describes the general items that can usually be found at the equipment level. The point is emphasized strongly that, at present, each piece of equipment may well need its own individually tailored controller and there is a need for further definition of the equipment level controller in order to permit a more generic design.

What is meant by equipment?

The automated manufacturing equipment resident at the equipment level can take a variety of forms but in general it will include the following categories.

ROBOTS

At the simplest level are the robots which learn a particular sequence of movements in 'teach' mode; ie the operator physically takes the robot through the sequence and stores this sequence in the robot controller. More sophisticated robots can learn a particular sequence directly through a communication link in a process called off-line programming, although there are many problems to be overcome in terms of the error involved. This error arises from the inherent backlash and other looseness in a robot joint; when commands are given to send a robot end effector to a particular position, there is a difference between the desired position and the actual position. The importance of this error will, of course, depend on the application. The real potential of off-line programming will perhaps only be fulfilled when an effective and relatively inexpensive mechanism is provided for accurately measuring the actual end effector position. If the end effector position can be accurately measured, then the error can be determined and the end effector moved to eliminate the error.

MATERIAL HANDLING DEVICES

This category includes those devices (other than robots) used to make movements within each cell, for example, the use of a small conveyor to move parts from an automated guided vehicle (AGV) to a machine.

TRANSPORT DEVICES

This category includes those devices used to make movements of parts *between* cells as compared with material handling devices which make movements *within* cells (although, of course, the division between categories is rather arbitrary). These transport devices include AGVs, conveyors and shuttle systems. The AGV systems can usually rely on sensing direction by some mechanism, the most common of which is the cable in or on the floor carrying a signal which is then picked up by a sensor on the AGV. Other sensory input include the use of coded visual markings on the floor; the AGV then uses ultraviolet (UV) light transmitters and receivers to determine the position. Advantages of the UV light approach are that the visual marking on the floor can be readily moved, and that the need for relatively sophisticated signalling processing (necessary for the buried wire approach) is eliminated. However, there may be some disadvantages to the UV approach. First, the markings can easily be disfigured, if, for example, the AGV path is used frequently by workers. Second, dynamic changes in AGV routeings cannot be made since there is no mechanism for altering the routeings Further developments in radio control of AGVs may make this possible, in that UV sensing is used as a position determinant and the radio control is used to instruct the AGV on which path to follow. The other major approach, the buried wire approach, is extremely popular. The buried wire can, instead of being buried, be taped to the floor surface, so that the path followed by the AGVs can be rapidly altered. Furthermore, the signal in the wire can be coded so that it gives instructions to the AGV on which route to follow. One of the latest developments is in the feedback through the wire of the actual AGV position, enabling better decisions to be made on AGV routeing. The buried wire approach therefore has an advantage in the ability to dynamically control the AGV route. One possible disadvantage is the added complexity of the signal processing necessary. The buried wires may, in addition, cause interference with other data communications, although this risk can be minimized by having some distance between the respective wires.

MACHINES

This category includes the NC machine tools, as well as other manufacturing machines such as electronic component insertion devices. The NC machine tools are largely responsible for metal cutting operations in those industries where this is the primary area. However, an increasing number of manufacturing systems involve electronic assembly or electronic component fabrication. For nearly all manufacturing systems, the upper echelons of the control hierarchy can be of basically the same format, with factory, shop and cell control being generically similar. The major difference appears at the equipment level, where the co-ordination of totally different tasks may require a different structure.

HUMAN RESOURCES

Most of what are loosely termed *automated* manufacturing systems require significant portions of human input at certain points. Full automation is difficult to achieve for a number of reasons, the major one being the occurrence of events such as tool breakage, machine breakdown, work being out of tolerance, etc. The adoption of sufficient automated procedures to handle all the possible events can be prohibitively expensive; the simplest solution is to rely on some human observer to monitor the system and to take action when required. Human labour may also be required in such areas as tool setting, inspection or jig/fixture loading. If human labour is to be used for these roles, then some provision has to be made for communication between them, the CCSs and the equipment level controllers (ELC).

The above categories of the equipment level are not meant to be exhaustive since different industries may have different equipment (for example, wood processing may require some chemical operation). For many industries, especially in aerospace, inspection processes may be important and a separate inspection cell or cells may be provided.

Equipment level control structure

The ELC takes as input, commands or goals from the CCS and translates these into commands for specific parts of the equipment. For example, a robot ELC receiving a command to move from position x to position y, will analyse the command and break it down into narrower commands for those joints, etc affected, and will co-ordinate the carrying-out of these narrow commands. The structure of the ELC will vary from application to application, but one likely future development is in increased ELC intelligence. With increased intelligence, decision making can be further decentralized to allow individual items at the equipment level to make decisions. This will have two major repercussions. First, instead of commands being issued from the CCS to the ELC only broad goals will need to be given. For example, it may well be possible to give a goal of the form:

move item xyz from buffer store 3 to machine C at time 11:37

to a particular robot. The robot ELC can then access its memory or calculate the robot program necessary to move the part, and at the desired time go ahead and move the part, contacting the CCS at the end of the cycle, or if there are any major problems associated with the goal set by the CCS. The second major repercussion is that some problem solving will be able to be done at the equipment level so that if, for example, a part is misplaced in a robot gripper, the robot ELC can decide what can be done to rectify the problem, again, only notifying the CCS if the problem cannot be rectified by the ELC.

The role of the CCS can therefore be reduced if sufficient intelligence and decision making can be incorporated into the ELC. The function of the CCS can be further reduced if there is a mechanism provided for inter-ELC communication, in that the sequential instructions can be passed from ELC to ELC. For example, if it is required that a robot move a part from a machine tool when the machine tool finishes work on that part, the machine tool ELC can notify the robot directly that it has finished work on the part and the robot can then move the part from the machine tool. In this manner the CCS is relieved of some of its tasks. This procedure does, however, require some mechanism to be provided for inter-ELC communication; this is usually taken to be some communication network within the cell.

As indicated earlier, a wide variety of automated manufacturing systems can have generically similar factory, shop and cell control systems; but there is a major difference in the equipment level controllers, in that the commands or goals set by the CCSs have to be broken down into specific instructions (for example, to elements such as servos within the particular equipment). Each piece of equipment will therefore usually require an individually tailored ELC.

Conclusion

This chapter has presented a brief overview of the ELC which forms the lowest level in the hierarchy of control proposed in earlier chapters. Each piece of equipment at the equipment level will usually have its own ELC, although it may be that some circumstances will dictate that one ELC should control more than one piece of equipment. The main commands or goals for the ELC are received from the CCS and the CCS usually acts as the co-ordinator of cell activities. The ELC is responsible for the detailed operation of its particular equipment.

The equipment found at the equipment level for automated manufacturing systems includes robots, material handling devices, transport devices, machines and human resources, although certain automated manufacturing systems may include more specialized equipment (such as automated inspection machines).

A wide variety of automated manufacturing systems can have similar factory, shop and cell control systems. However, the varying equipment found at the equipment level will usually mean that, at present, each ELC can be used on only that type of equipment.

CHAPTER 9

Conclusion and Future Trends

Production planning and control for a batch manufacturing system usually involves the simultaneous consideration of a large number of product types flowing through a number of different processing machines. The determination of effective planning and control can be a very time consuming and difficult process. When dealing with an automated manufacturing system, there are a number of factors which make production planning and control problems even more complex.

This book has presented the background to production planning and control of automated manufacturing systems and various approaches to the problem have been discussed. The presentation of material in the book has intended first, to give an overview of production planning and control, and its application to automated manufacturing systems; second, to discuss hierarchical approaches; and finally, to progress sequentially down the hierarchy describing each level.

Chapter 1 detailed some definitions of automated manufacturing systems that have been proposed, and the term 'flexibility' was discussed with reference to automated manufacturing systems. The increased importance of production planning and control for automated manufacturing systems in comparison with conventional manufacturing systems was also stressed, and this increased importance was suggested as due to two major factors. First, since it is important for lead times to be extremely low in automated manufacturing systems, activities must be scheduled and controlled more closely to achieve these reduced times. It may well be, of course, that where there is a conventional manufacturing system with extremely short lead times, then it may also require a similar detailed production planning and control structure. However, in conventional manufacturing systems, there is usually sufficient slack for detailed production planning and control not to be needed.

The second major reason for the increased importance of production planning and control for automated manufacturing systems is that a typical automated system has an extremely high capital cost, so high usage becomes very desirable. Simultaneous achievement of high usage and other desirable attributes, such as short lead times, low work-in-progress and due date achievement, relies on an effective production planning and control system.

The particular characteristics of automated manufacturing systems and their impact on production planning and control were discussed further in Chapter 2. Some aspects of automated manufacturing systems make the production planning and control problem somewhat easier. These aspects of production planning and control in conventional manufacturing systems were described in Chapter 3. The approach described, which is what may be termed the traditional approach, broke the problem down into three levels in a hierarchy of planning and control: long term, medium term, and short term levels. In Chapter 3, major attention was focused on the latter two levels, since the production planning and control function is concentrated at these levels. An overview was therefore given of such traditional approaches as master production scheduling, materials requirements planning and job shop scheduling and the role and limitations of each were considered.

In Chapter 4, two approaches to the hierarchical control of automated manufacturing systems were described, these two approaches being the advanced factory management system produced by Computer Aided Manufacturing Inc, and the control structure of the automated manufacturing research facility produced by the National Bureau of Standards. From these two approaches a generic control hierarchy was presented. This generic control hierarchy contains four levels: factory, shop, cell and equipment levels. Each higher level receives feedback from the level directly below it and the higher level has control over this lower level. Each of these levels was described in more depth in Chapters 5 to 8.

The factory level control was described in Chapter 5.

This level is concerned with overall factory computer and information systems, and it forms the highest level in the proposed control hierarchy. Within this level are software modules such as financial systems, computer aided design, process planning, master production scheduling I and materials requirements planning. The output to the next lowest level, the shop level, and to other lower levels usually consists of requirements for work to be completed by the automated manufacturing system (sometimes termed the 'work-to' list) as well as process planning information. Since the factory level control forms the highest level of the control hierarchy, managerial interaction will have the greatest effect at this level; for this reason, this level probably constitutes the point at which the majority of managerial interaction occurs.

Most of the output from the factory level control will pass directly to the shop level and this level was discussed in Chapter 6. The shop control system (SCS) which resides at shop level takes requirements and data (in the form of a 'work-to' list) and process planning information from the factory level, as well as feedback from the cell level. The SCS then produces commands or goals for each cell control system (CCS) at the cell level, with the amount of detail required in the commands or goals depending on the decision making abilities at the shop and cell levels. Within the SCS, then, two levels can be readily identified. The first level, master production scheduling II, extracts a viable work load, for a particular time period of typically a few hours, from the 'work-to' list. This MPS II function takes into account such factors as due dates as well as tooling and other constraints. The output from the MPS II passes to on-line scheduling, where a more detailed schedule of activities within each cell is produced. Two major approaches to MPS II and on-line scheduling were presented. These are an adapted goal programming formulation for MPS II and adaptive heuristics for on-line scheduling. However, these two approaches can be augmented by other approaches arranged in a 'mailbox' format, in order to arrive at good decisions in relatively short computing times.

The goals or commands pass from the shop level to the cell level, and Chapter 7 described the CCS. The CCS takes goals or commands from the shop level and translates these into specific cell activities. The degree of decision making and autonomy possible at cell level depends on the intelligence and memory available to the CCS and four categories of CCS were proposed ranging from Class IV low intelligence, low memory to Class I high intelligence, high memory. The typical constituents of a cell were then discussed, with the definition of a cell ranging from a simple processing unit containing a single machine to the virtual manufacturing cells of the automated manufacturing research facility, which contain all the machines necessary to completely process a product family. The modes of operation of the CCS were divided into four: highly centralized, loading, itemized, and decentralized. Each of these modes was described and the advantages and disadvantages of each discussed.

The lowest level of control in the proposed production planning and control hierarchy is at the equipment level, and this was discussed in Chapter 8. The data input from the higher levels to the equipment level controller (ELC) consists of specific commands from the CCS, the CCS acting as a co-ordinator of cell activities. Each piece of equipment at the equipment level will usually have its own ELC and each ELC is responsible for the detailed operation of its particular equipment. The major categories of equipment were described and the relationship between the ELC and the CCS was discussed.

Overall production planning and control functions

The structure of the production planning and control system for automated manufacturing systems was described in Chapters 4 to 8, and the major production planning and control functions are as shown in Figure 9.1. The functions include:

1. *Master production scheduling I (MPS I).* This is the overall corporate MPS I, where the production rate of each end product or product family is determined, usually for each week, over the planning horizon (which is often one year).
2. *Materials requirements planning (MRP).* For assembly-type industries, output from the MPS I is fed into the MRP system to give net requirements, in the form of a work-to list for materials and components. There may be some element of capacity balancing at this stage.

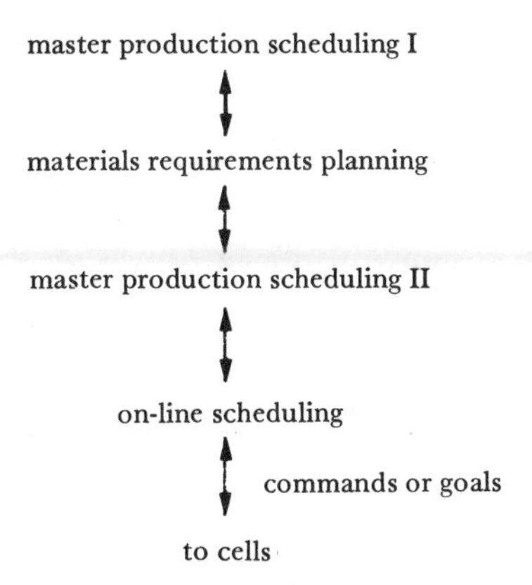

Figure 9.1 *Major production planning and control functions for automated manufacturing systems*

3. *Master production scheduling II (MPS II).* The work-to list is passed down to the MPS II module. The function of this module is to extract from the work-to list a viable subset of jobs to process in the next time period (usually one day, or one shift of eight hours or so). This function is necessary where there are detailed constraints such as tooling, jigs/fixtures or transport. A suitable methodology for achieving this using an adapted goal programming formulation was presented.

4. *On-line scheduling.* The viable subset of jobs produced by the MPS II module is now scheduled on to each machine or process. A suitable methodology using an adaptive heuristic was presented.

It should be noted that the above hierarchy is a flexible one – each function can be added to or deleted as and where necessary. So that, for example, where there are no detailed engineering or other constraints, the MPS II function can be left out. Similarly, where the automated manufacturing system is in a non-assembly type industry, the MRP module can be deleted.

The MPS I and MRP modules are conventional: they are used in both automated environments as well as in the more traditional manual systems. The MPS II and on-line scheduling modules are specifically designed for automated manufacturing systems. Future developments in such emerging technologies as artificial intelligence (AI) and expert systems could significantly add to the modules presented in this book, as described below.

Future trends

There are a number of trends that may have a substantial impact on production planning and control of automated manufacturing systems in the future. Two major developmental areas are first, in the use of artificial intelligence as an aid to decision making and second, in the use of what may be termed 'Japanese' approaches to production management. Each of these will be briefly described.

The term artificial intelligence can be rather elusive to define strictly. Broadly speaking, it encompasses such areas as expert systems, knowledge representation, distributed problem solving and knowledge management. There is a considerable amount of work being carried out, mostly in USA, Japan and Europe, into artificial intelligence and it is to be expected that some of the developments in this area will have an impact on production planning and control. Perhaps the most mature area of artificial intelligence is that of expert systems, where an expert's knowledge of a particular field is captured by a series of rules in what is called the expert rule base. An inference engine is then used to extract knowledge from the expert rule base so that it can be of help to a user. The most conventional approach to the inference engine and expert rule base is to use *production rules*, where knowledge is required in a series of IF-THEN rules, for example:

IF machine 32 is overloaded THEN route parts through machine 78

On each pass through the production rules, some are satisfied; these satisfied production rules form the basis of suggested actions. The application of expert systems to production planning and control of automated manufacturing systems is probably limited to those areas where an expert could be useful. Such areas include process planning, CAD, the mailbox format in the SCS and the evaluation of sensory feedback at the equipment level. However, the application of other artificial intelligence techniques is likely to have far-reaching effects on production planning and control in the future.

The 'Japanese' approach to production planning and control has a number of points of emphasis. The first is to break the whole system into a number of smaller, more easily managed subsystems. The second is to arrange production so that lead times and work-in-progress levels are reduced while quality levels are increased. Considering the first point, the layout of the batch manufacturing facility has historically been on a process basis, where similar process operations are lumped together in one area; so a typical batch manufacturing concern may consist of an area set aside for lathes, another for milling machines another for grinding machines, and so on. A job going through such a shop, therefore, may well have to move some considerable distance in its progress through the shop. Recent studies of 'Japanese' type approaches to production planning and control and just-in-time systems has focused some attention on layout (as well as on a number of other aspects) and the result has been a renewed interest in moving towards small flow lines. These small flow lines are based on breaking down the product range produced by a manufacturing concern into a number of generically similar product families, each product family being produced on a flow line. In this manner, the inefficiency associated with the process-type layout can be reduced and, in addition, these small flow lines can be easier to control than the larger scale manufacturing facility. The second aspect can be approached by a systematic lowering of work-in-progress levels. As work-in-progress is lowered, so problems may be encountered. The basis of the Japanese approach is to remove these problems as and when they occur, rather than to raise work-in-progress levels to cover the problem areas. In this manner, for example, if a particular machine has a poor reliability, sufficient work-in-progress could be kept to ensure that subsequent machines do not run out of work if that machine breaks down. The Japanese approach would be to arrange activities via, perhaps, a preventive maintenance programme to avert a machine breakdown, thereby allowing work-in-progress levels to be reduced. Similar actions could be performed on traditional problem areas such as supplier performance and quality control.

The major repercussion felt by the Japanese approach to automated manufacturing systems is in the emphasis on flow lines, lead time reduction and high quality levels. Such considerations are also the major factors in encouraging investment in automated manufacturing systems; automated manufacturing systems can therefore be considered as a realization of the Japanese and just-in-time approaches.

Conclusion

This book has presented a realistic and viable plan of attack on the problem of production planning and control in automated manufacturing systems. A structure has been described which ties in with existing production engineering and production management approaches, which may already be firmly established in a manufacturing concern. The approaches that have been described are adaptable to a wide variety of system designs and they have been tested on a number of practical manufacturing systems. The importance of an effective production planning and control system for automated manufacturing systems cannot be overestimated: such a production planning and control system can result in a much improved operation of the expensive automated facility.

Appendix I

Master Production Scheduling II

The mathematical formulation involved in the approach of O'Grady and Menon (1985) is described in this appendix.

The weighted attainment function

$$Z(d) = \sum_{m=1}^{t} (w_m^+ d_m^+ + w_m^- d_m^-)$$

where $Z(d)$ is the attainment function

$d_m^+ (d_m^-)$ is the deviation variable monitory overachievement (underachievement) of target m

w_m^+, w_m^- are the weights for deviations (d_m^+, d_m^-) to reflect relative preferences in the attainment function

The individual modules, which can be used to build up the model to suit the user's requirements, are described below.

1. *Tooling requirements*

   The restraint to achieve this aspect is:

   $$x_{ijk} - y_{jk} \leqslant 0 \quad (\forall_i \in I, \forall_j \in J, \forall_k \in K)$$

   where:

   $x_{ijk}(n)$ is the zero-one variable for candidate order i requiring special tool k at machine j with n optional process routes

   y_{jk} is the zero-one variable of candidate tool k for assignment to machine j

   $\forall$ for all

   $\in$ is a member of the set.

2. *Linked groups of orders*

   The set $S = x_{i1}, x_{i2}, x_{i3}, \ldots, x_{ig}$ is the set of orders that are to be linked. The restraints to specify this linking are:

$$x_{i1} - x_{i2} = 0$$

$$x_{i2} - x_{i3} = 0$$

$$x_{ig-1} - x_{ig} = 0$$

3. *Tool magazine capacity*
   The restraint to deal with tool magazine capacity problem are of of the form:

$$\left[\sum_{k=1}^{r} y_{jk}\right] + ST_j - d_{ij}^{+} + d_{ij}^{-} = MC_j\,(\forall_j \in J)$$

   where: r is the number of tools and ST_j is the standard tooling partition of magazine at machine j.

4. *Tool type availability*
   These restraints are in the form:

$$\left[\sum_{j=1}^{q} y_{jk}\right] - d_{2k}^{+} + d_{2k}^{-} = TT_k\,(\forall k \in K)$$

   where: q is the number of machines and TT_k is the number of tool type k currently available.

5. *Machine capacity*
   The nominal machine hours per machine MH_j can be given with penalties for under or overachieving this figure with the restraints of the form:

$$\left[\sum_{i=1}^{p} h_{ij}\, x_{ij}\right] - d_{3j}^{+} + d_{3j}^{-} = MH_j\,(\forall_j \in J)$$

   where: h_{ij} is the number of hours required by order i at machine j and MH_j is the number of machine hours available at machine j.

6. *Alternative process routes*
   The selection of a maximum of one route is ensured by the restraint:

$$\sum_{n=1}^{u} x_{ijkn} \leqslant 1$$

   where: u is the number of alternative process routes for order i.

7. *Due date consideration*
   The restraints necessary to discriminate in favour of overdue orders are; for overdue orders:

$$\left[\sum_{i=1}^{p} b_i\, x_i\right] - d_4^{-} = 0\,(\forall_i \in I \mid b_i < 0)$$

For orders within due date:

$$\left[\sum_{i=1}^{p} b_i x_i\right] - d^+ = 0 \ (\forall_i \in I \mid b_i \geqslant 0)$$

where: b_i is the due date coefficient for order i which indicates time remaining (> 0) or overdue (< 0) with respect to the promised delivery date, and p is the number of orders.

8. *Product release level*
   The amount of products released can be controlled by the restraints:

$$\left[\sum_{i=1}^{p} c_{i1} x_i\right] - d_5^+ + d_5^- = RL_i \ (\forall 1 \in L)$$

where: c_{i1} is the quantity of attribute for order i

9. *Expediting certain orders*
   A priority index, e_i where $e_i > 0$, can be chosen by the manager to give preferential consideration to order i, with the restraint being of the form:

$$\left[\sum_{i=1}^{p} e_i x_i\right] - d_6^+ = 0$$

A 'hard' constraint formulation is:

$$x_i = 1 \ (i \subseteq I \mid e_i > 0)$$

References

Bell, R.; Bilalis, N. (1982) *Loading and Control Strategies for a FMS for rotational parts* 1st Int. Conf. on FMS, October, IFS Publications.

Bergstrom, G.L.; Smith, B.E. (1970) Multi-item production – an extension of the HMMS rules. *Management Science* **16**, B614-629.

Bloom, H.M.; Furlani, C.M.; Barbera, A.J. (1984) Emulation as a design tool in the development of real-time control systems. *Proceedings of Winter Simulation Conference* November 28-30 1984, Dallas.

Buzacott, J.A.; Shanthikumar, J.G. (1980) Models for understanding flexible manufacturing systems. *AIIE Trans.* **12**, No. 4, December 1980.

Carrie, A.S.; Adhami, E.; Stephens, A.; Murdoch, I.C. (1983) Introducing a flexible manufacturing system. *Proc. Seventh International Conference on Production Research* Windsor, Canada.

Chang, R.H.; Jones, C.M. (1970) Production and workforce scheduling extensions. *AIIE Transactions* **2**, p. 326.

Cheng, E.T.C. (1982) A combined approach to due date scheduling. In *Advances in Production Management Systems '82* pp. 295-303. International Federation for Information Processing Conference, Bordeaux, 1982.

Draper Laboratories, Charles Stark (1983) *Flexible Manufacturing System Handbook* **I-IV**. U.S. Dept. of Commerce. AD/A127, pp. 927-930, February 1983, NTIS Publications.

Elmaleh, J.; Eilon, S. (1974) A new approach to production smoothing. *International Journal of Production Research* **12**, p. 673.

Eversheim, W.; Fromm, W. (1983) Production control in highly automated manufacturing systems. *Proc. AUTOFACT Europe Conf.* September, pp. 3-1–3-13, Geneva.

Fischer, H.; Thompson, G.L. (1963) Probabilistic learning combinations of local job shop scheduling rules. Chapter 15 of *Industrial Scheduling* (J. Muth; G. Thompson eds.) Prentice Hall, New Jersey.

Fox, K. (1982) Simulation for design and scheduling for flexible manufacturing systems. *Proc. AUTOFACT* **4**, pp. 6-27–6-36, December, CASA/SME.

Furlani, C.M.; Kent, E.W.; Bloom, H.M.; McLean, C.R. (1983) *The Automated Manufacturing Research Facility of the National Bureau of Standards* Proc. Summer Computer Simulation Conference July 11-13 1983. Vancouver, B.C., Canada.

Groover, M.P. (1980) *Automation, Production Systems and Computer-Aided Manufacturing* Prentice-Hall, Englewood Cliffs, N.J.

Gunther, H.O. (1981) *A Comparison of Two Classes of Aggregate Production Planning Models under Stochastic Demand* Second International Working Seminar on Production Economics, February 16-20, 1981, Innsbruck.

Hershauer, J.C.; Ebert, R.J. (1975) Search and simulation of a job-shop scheduling rule. *Management Science*, **21**, pp. 833-843.

References

Holt, C.C.; Modigliani, F.; Muth, J.; Simon, H.A. (1960) *Planning Production, Inventories and Work Force* Prentice Hall, New Jersey.

Ingersoll Engineers (1982) *The FMS report* IFS Publications.

Jones, A.T.; McLean, C.R. (1984) A cell control for the AMRF. *ASME Conf.* August 1984.

McLean, C.; Bloom, H.; Hopp, T. *The Virtual Manufacturing Cell* IFAC/IFIP Conference on Information Control Problems in Manufacturing Technology. October 1982. Gaithersburg, MD.

O'Grady, P.J. (1981) *The Application of Discrete Modern Control Theory to the Problem of Production Planning and Control* Ph.D. Thesis, University of Nottingham.

O'Grady, P.J.; Byrne, M.D. (1985) A combined switching algorithm and linear decision rule approach to production planning. *International Journal of Production Research*, **23**, pp. 285-296.

O'Grady, P.J.; Harrison, C. (1985) A search sequencing rule for job shop sequencing. To be published in *International Journal of Production Research.*

O'Grady, P.J.; Menon, U. (1985) A multiple criteria approach for production planning of automated manufacturing. *Engineering Optimization* **8**, pp. 161-175.

O'Grady, P.J.; Menon, U. (1986) *Master Scheduling for a Flexible Manufacturing System with Tooling Constraints: A Case Study*. To be published.

Orlicky, J. (1975) *Materials Requirements Planning* McGraw-Hill.

Orr, D. (1962) A random walk production-inventory policy: rationale and implementation. *Management Science* **9**, p. 108.

Panwalker, S.S.; Smith, M.L.; Seidmann, A. (1982) Common due date assignment to minimize total penalty for the one machine scheduling problem. *Operations Research* **30**, pp. 391-399.

Rochette, R.; Sadowski, R.P. (1976) A statistical comparison of the performance of simple dispatching rules for a particular set of job shops. *International Journal of Production Research* **14**, pp. 63-75.

Solberg, J. (1976) Optimal design and control of computerized manufacturing systems. *AIIE Conf.* pp. 138-147, Boston.

Stecke, K.E. (1983) Formulation and solution of nonlinear integer production planning problems for flexible manufacturing systems. *Management Science*, **29**, pp. 273-288.

Welam, U.P. (1975) Multi-item production smoothing models with almost closed form solutions. *Management Science* **21**, p. 1021.

Yao, D.D.W. (1983) *Queueing Models of Flexible Manufacturing Systems* Ph.D. Thesis, University of Toronto.

Index